Residential Open Building

Also available from E & FN Spon

Building International Construction Alliances
R. Pietroforte

Construction - Craft to Industry
G. Sebestyen

Construction Methods and Planning
J.R. Illingworth

Creating the Built Environment
L. Holes

Green Building Handbook
T. Woolly, S. Kimmins, R. Harrison and P. Harrison

Industrialized and Automatic Building Systems
A. Warszawski

Introduction to Eurocode 2
D. Beckett and A. Alexandrou

Open and Industrialised Building
A. Sarja

The Idea of Building
S. Groak

Procurement Systems
Edited by S. Rowlinson and P. McDermott

Journal
Building Research and Information
The International Journal of Research, Development, Demonstration & Innovation

To order or obtain further information on any of the above or receive a full catalogue please contact:
The Marketing Department, E & FN Spon, 11 New Fetter Lane, London EC4P 4EE
Tel: 0171 842 2400; Fax: 0171 842 2303

Residential Open Building

Stephen Kendall
Housing Futures Institute, Ball State University
and
Jonathan Teicher
Building Community, The American Institute of Architects

Routledge
OPEN BUILDING SERIES

London and New York

CRC Press
Taylor & Francis Group
6000 Broken Sound Parkway NW, Suite 300
Boca Raton, FL 33487-2742

First issued in paperback 2019

ISBN-13: 978-0-419-23830-0 (hbk)
ISBN-13: 978-0-367-39898-9 (pbk)

British Library Cataloguing in Publication Data
A catalogue record for this book is available from the British Library

Library of Congress Cataloging in Publication Data
Residential Open Building / Stephen Kendall & Jonathan Teicher
            p. cm.
    Includes bibliographical references and index.
    1. Dwellings--Design and construction   2. Open plan (Building)
I. Teicher, Jonathan.   II. Title
TH4812.K45   2000
690' .8--dc21
                                  99-33925
                                      CIP

Visit the Taylor & Francis Web site at
http://www.taylorandfrancis.com

and the CRC Press Web site at
http://www.crcpress.com

Contents

What is residential Open Building?

Throughout North America – and increasingly, throughout the world – non-residential buildings are constructed in an **Open Building (OB)** approach. Office and retail developers, their design and construction teams, and the associated regulators, lenders, owners, tenants, and product manufacturers are reorganizing the building process. They routinely work according to principles and methods that have developed over recent decades in direct response to extraordinary and accelerating change in the shaping of environment.

Regardless of style, typology or construction, commercial **base buildings** are now customarily built without predetermined interior layout. Upon leasing, demising walls and then interior partitioning are added, as spaces are fitted out to suit individual tenants. Each tenant may install unique interior spaces, equipment and systems to suit organizational and technical needs. When older commercial buildings are 'revalued,' demolition exposes the existing building shell, which is then retrofitted with upgraded facade and interior systems. Even in 'build-to-suit' office facilities, base building construction is made as generic as possible: its long-term value is increased by providing capacity for changing requirements, including eventual tenant turnover and future sale.

Developments in commercial construction are now moving into the residential sector. In Europe, Asia and North America, **residential Open Building** principles, variously known as *OB, S/I (Support/Infill), Skeleton Housing, Supports and Detachables, Houses that Grow,* etc. – are now spearheading the reorganization of the design and construction of residential buildings in parallel ways. In many cases, residential Open Building is based on the reintroduction of principles intrinsic to sustainable historic environments around the world. These have been reinterpreted and updated to harness benefits of state-of-the-art industrial production, emerging information technologies, improved logistics, and changing social values and market structures.

Residential Open Building is a new multi-disciplinary approach to the design, financing, construction, fit-out and long-term management processes of residential buildings, including mixed-use structures. Its goals include creating varied, fine-grained and sustainable environment, and increasing individual choice and responsibility within it. In Open Building, responsibility for decision-making is distributed on various levels. New product interfaces and new permitting and inspection processes disentangle subsystems toward the ends of simplifying construction, reducing conflict, affording individual choice, and promoting overall environmental coherence. Residential OB thus combines a set of technical tools with a deliberate social stance toward environmental intervention.

Residential Open Building practices are rapidly evolving throughout the world. As new consumer-oriented infill systems appear and become more widely available, governments, housing and finance corporations and manufacturers are joining developers, sustainability advocates and academics in endorsing and advancing a new open architecture. From improved decision-making and increased choice, to standardized interfaces between building systems that are compatible and sustainable, the broadly-shared benefits of the 'new wave in building' (Proveniers and Fassbinder, n.d.) are increasingly in evidence throughout the world.

Acknowledgments

The word in language is half someone else's. It becomes 'one's own' only when the speaker populates it . . . Prior to this moment of appropriation, the word does not exist in a neutral and impersonal language (it is not, after all, out of a dictionary that a speaker gets words!) but rather it exists in other people's mouths, in other people's concrete texts, serving other peoples' intentions . . .
– M.M. BAKHTIN, *THE DIALOGIC IMAGINATION*

Residential Open Building practice continues to grow. Groups and individuals far too numerous to mention have contributed to residential OB as it has taken root throughout the world. They have built projects, funded and conducted research, written, taught and organized in support of an open architecture. This book, stimulated by the formation of an international group focussed on the implementation of Open Building, is the first to chart the world-wide developments. It lends its voice to join in an ongoing storytelling process, toward the end of enriching the practice of a people-centered open architecture and the reorganization of the building process.

In the architecture and allied building industries of many nations, formal and deliberate Open Building in project development, design and implementation, in product manufacture, construction and systems installation, and in the management of real estate assets, has unevenly entered the mainstream of discourse and practice. Like the term 'Open Building,' the movement's pioneers, practices and principles sometimes remain unknown. Regrettably, this book can acknowledge only some of the trailblazers; and it presents only some realized projects, although Open Building has advanced no less through countless additional research efforts and unrealized projects, particularly at the level of urban tissues.

This introductory survey of the history, principles, and worldwide state of the art of Open Building is a direct outgrowth of the early meetings of the CIB's Task Group 26 Open Building Implementation. Without substantial support from the CIB Secretariat and TG 26 members and supporters, it could not have come about. We are particularly grateful for the efforts of Wim Bakens, Secretary General of the CIB, who remains a long-term and enthusiastic supporter of Open Building implementation. While sole responsibility for opinions expressed in this book rests with the authors, many other Task Group members contributed directly and substantially to the assembly of survey information on which this primer reports. Their invaluable and ongoing contribution is gratefully acknowledged here, as well as in the Parts that follow. To Ulpu Tiuri of Finland, Ype Cuperus and Karel Dekker of the Netherlands and to Seiichi Fukao, Shinichi Chikazumi, Hideki Kobayashi, Mitsuo Takada, Seiji Kobata and Kazuo Kamata of Japan, special thanks are due for providing materials and information at crucial points throughout the writing of this book. Without the invaluable contribution of book designer Ori Kometani and production assistant Jennifer Wrobleski, and the support of our families and publisher, and of Janet R. White, FAIA, this book could not have been published. More generally, we join worldwide supporters of OB in being indebted to Yositika Utida and Kazuo Tatsumi in Japan, and Age van Randen in the Netherlands. They have, among many other pioneers, been leaders in the unfolding story of Open Building.

Of all contributors to OB, perhaps none has had greater influence worldwide, over a number of decades, than N.J. Habraken, author of books ranging from *Supports: An Alternative to Mass Housing* (1961) to *Variations: The Systematic Design of Supports* (1976) to *The Structure of the Ordinary* (1998); Founding Director of the SAR research foundation and of the Department of Architecture at Eindhoven; former Chair of MIT's Department of Architecture; founding CEO of Infill Systems BV; co-inventor of the Matura Infill System; mentor and friend to so many advocates of an open architecture worldwide.

Unavoidably, indelibly, and with gratitude, this work echoes the profound influence and humanity – and many times, inevitably, the actual words – of John Habraken.

PART ONE

A RESIDENTIAL OPEN
BUILDING PRIMER

1

Introduction

1.1 THE OPEN BUILDING MOVEMENT

Developments toward residential Open Building are widespread and accelerating. They accompany change: in environmental structure, in production and construction methods, in the market for services and products, in product technology, and in the demand for suitable housing. However, unlike many new products or methods, Open Building was not invented. It has not developed in a unified fashion. Nor has it been aggressively marketed or promoted by multinational corporations, governments or associations. Rather, OB has emerged gradually in response to evolving social, political and market forces, to prevailing conditions and trends in residential construction and manufacturing, and to many other factors that demand more effective and responsive practices.

Developments toward residential Open Building respond to many of the same long-term environmental and social shifts that have affected non-residential architecture. Realized OB projects build on concerted long-term research, development and implementation activities conducted by individuals, corporations, associations, industries and government agencies. Yet even more, these projects and research activities reflect direct advocacy – of consumer choice and tenants' rights, of rationalized production of housing and building systems, of long-term environmental coherence or of sustainable architecture.

Parallel trends emerging across professions and regions have taken decades to recognize. More continue to appear. Residential Open Building advocacy now constitutes an international phenomenon. Gradually, groups and individuals – from industrial component manufacturers to real estate developers and contractors, from tenants' rights advocates to architects, and from sustainability advocates to government regulators – have come to recognize that they face similar problems, share similar beliefs about how to build (though frequently for different reasons), and have developed parallel or complementary – albeit not identical – responses to similar conditions. Above all, they understand that buildings are built and maintained through the concerted efforts of many parties operating at many different levels. It therefore makes sense to structure the interfaces of parts and of decision-makers in ways that improve the responsiveness of buildings to end users, while at the same time increasing efficiency, sustainability and capacity for change, and dramatically extending the useful lives of residential buildings.

1.2 TRENDS TOWARD OPEN BUILDING

The broadest environmental trend leading professionals toward Open Building practice is the reemergence of a changeable and user-responsive **infill (fit-out) level**. Infill represents a relatively mutable part of the building. The infill may be determined or altered for each individual household or tenant without affecting the **Support** or **base building**, which is the building's shared infrastructure of spaces and built form. Infill is more durable and stationary than furniture or finishes, but less durable than the base building.

Also noteworthy is the trend toward increasing building project complexity in terms of size, regulatory processes, systems coordination, production and management processes. Historically – until perhaps 75 years ago – patterns of residential development, decision-making, construction and control were relatively constant. Now they are rapidly shifting. As one result, any direct or substantive participation in decision-making by the end user or inhabitant is now frequently removed from the building process.

By contrast, within commercial office buildings, rights and responsibilities for selecting and maintaining major components of building and equipment subsystems is shifting toward the tenant. Investment is steadily moving to the fit-out and furniture levels – where it becomes the end users' personal property – rather than to the level of the base building – which constitutes real property with very different constraints and business drivers (Ventre, 1982). Building procurement and service subcontracting are rapidly evolving, differentiating and transforming to match these changes in investment patterns.

Many other broad environmental trends are aligned with developments toward residential Open Building. Among the technical trends, an increasing number of high-value-added subsystems are being introduced into buildings with ever-increasing frequency. Multiple and highly complex utility supply systems are being extended into every space within the home. Industrially-produced technical supply systems and building products increasingly proliferate, become physically entangled on-site and then become obsolete and abandoned, like piping for gas lamps or rooftop antenna leads.

In terms of project finance, the rate of investment in refurbishing and maintaining existing building stock is sharply rising. Renovation now accounts for more than half of the construction market in many developed nations. Yet the relative capacity of buildings to adapt to changes in infill systems, use or user preferences has greatly diminished. The average life span of new buildings has plummeted, from 100 years to as few as 20 or 30. Developers and contractors are also keenly aware of yet another long-term global trend: construction dollars are flowing away from site construction toward prefabricated (made for use) and industrially-produced (made for stock or trade) subsystems.

1.3 HOW OPEN BUILDING WORKS

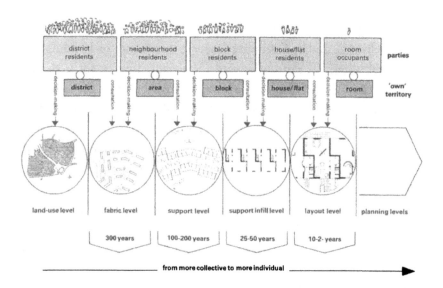

Fig. 1.1 Decision-Making Levels in Open Building. Diagram courtesy of Age van Randen.

Organizationally, Open Building lends formal structure to traditional and inherent levels of environmental decision-making, while offering design methods based on new insights and supported by current applied research. OB projects are structured to subdivide technical, aesthetic, financial and social decisions into distinct **levels** of decision-making. **Urban level** decisions address the wider public realm, including the establishment of urban patterns of built form and space, placement of streets, parking and utility networks, setbacks and 'street furniture.' They may further address the character of building facades, the location of public buildings, and the distribution of activities (land uses) within the more enduring spatial and formal order of the urban tissue.

Within that urban structure, independent decisions on the **Support (base building) level** involve the parts of a building which are common to all occupants, those parts which may endure for a century or more. To use multi-family housing as an illustration: the base building may be comprised of the load bearing structure, plus the building's common mechanical and conveyance systems and public areas, as well as all or most of its outer skin. Individual tenant changes can – and should – leave the Support unaffected.

Systems and parts associated with the **infill (fit-out) level** tend to change at cycles of 10-20 years. Transformation may be occasioned by occupants' changing requirements or preferences, by cyclical need for technical upgrade or by changes in the base building. The infill typically comprises all components specific to the dwelling unit: partitioning; kitchen and bathroom equipment and cabinets; unit heating, ventilating and air conditioning systems; outlets for power, communications and security; and all ducts, pipes and cables which individually service facilities in each unit. In detached houses, OB distinguishes changeable interior fit-out from more durable structure and skin.

In open architecture, these infill parts may be independently installed or upgraded for each occupant in turn. To make that possible, the base building must be kept as physically distinct as possible from its less permanent infill. To enable the independence of the infill, buildings cannot be built as single integrated 'bundles' of technical products or decisions. The separation intrinsic to an open architecture invests additional value, possibility and durability in the Support. Which is to say, the Support structure builds in valuable capacity for lower level change. Infill systems and parts will inevitably have to be changed many times throughout the life cycle of the building in which they are located. Therefore, they are designed and installed for optimal freedom of independent layout, construction, subsequent transformation and eventual replacement. At the same time, common systems and long-term durable parts shared by all occupants – for instance, foundations, structure, utility trunk lines, public corridors and stairs – are left viable and undisturbed.

In terms of **decision clusters**, Open Building therefore advocates **disentangling** specific parts of buildings and their sub-assemblies: minimizing interference and conflict between subsystems and the parties controlling them; and enabling the substitution or replacement of each part during design, construction and long-term management. These principles apply to work at each environmental level. They also apply to both residential and non-residential architecture. Disentangling and standardizing interfaces in residential and commercial architecture alike enables broad consumer choice in laying out, equipping and finishing spaces. The use of residential fit-out systems has begun to restructure residential construction, which is consequently emerging as a new kind of consumer market.

In one current example of state-of-the-art Open Building practice, households work with an infill architect to custom design their own dwelling units according to their functional, aesthetic, budget and other preferences. The future inhabitants decide where to place walls, kitchen and bathroom. They select cabinets, appliances, fixtures and finishes. A few weeks later, they can move into their newly constructed and code-approved custom dwellings. The consumer-grade utility systems and their connections and interfaces, rather than the particular installation, have received code-approved product certification, streamlining local jurisdiction inspections. As one result, occupants are free to subsequently relocate electrical, data and communications outlets at will. Such entirely custom dwellings, using advanced infill technology, information systems and logistics, need not cost more than conventional units.

Principles and practices such as these are transforming conventional practice in urban design, architecture and construction. They are also re-shaping the processes of designing, manufacturing and installing building subsystems and parts. New processes and forms of organization in the design and construction professions are taking shape around OB, and new building technologies and materials are being produced to suit it. Building standards, regulations, financing and management are adjusting in ways compatible with Open Building practice.

2

Incubators of Open Building

2.1 THE NETHERLANDS

2.1.1 John Habraken and *Supports*

In 1961, N.J. Habraken, a young Dutch architect, published *De Dragers en de Mensen: het einde van de massawoningbouw,* a slender volume subsequently translated as *Supports: An Alternative to Mass Housing* (1972). In *Supports,* Habraken observed that mass housing (MH) had begun to disrupt an age-old 'natural relation' between human beings and their built environment. Although the brutalist forms of mass housing might be embraced or villified, there was far more at stake than style: as one by-product of the reorganization of the housing process, mass housing was creating a disruptive imbalance among forces which, in healthy environment, operate in dynamic equilibrium. Largely implicit processes had hitherto created, sustained and enriched built environment for millenia, based on slowly evolving themes and variations. Now those processes were being brought to a halt.

In the traditional process of habitation, each household had of necessity acted directly to take charge of the act of dwelling. Within a block of Amsterdam canal houses, for example, there existed a clear and common typology (and in fact, a collective urban structure of high coherence). Yet each inhabitant or owner independently managed and altered his or her own dwelling. Every stoop, every facade, every window and every plan was therefore different, a unique and vital variation of a broadly shared environmental theme. Mass housing had utterly excluded such participation and responsibility of individual households, entirely eliminating inhabitants from the housing process. In the new post-WWII building order, everything was professionally decided, professionally designed and professionally managed and maintained. Built environment as a collective artifact embodying people in all of their individuality and uniqueness was dying. Habraken perceived that dwellings could not be understood as products or manufactured objects. Dwelling was, rather, a fundamentally human process. And the issue was not aesthetics, nor even industrialization, but rather unified institutional control of what had been an activity shared collectively by members of society.

Habraken believed it was possible to reinstate the natural relationship or process within built environment. To restore healthy environmental structure to residential areas in the face of new and rapidly changing environmental conditions would require some form of support. Residents needed to be able to make autonomous dwelling decisions on their own behalf, rather than to be provided with units of housing (Habraken, 1999). They also needed to live in stacked multi-family dwellings, where they could somehow 'plug in' to multiple supply systems. They would nestle within a three-dimensional structure; nonetheless, they would remain free to transform their homes. Inhabitants would remain unaffected by reverberating change initiated by neighbors, including renovation or even total demoli-

Living is an act that takes place in both spheres.
A home connects the two spheres:
A home is the environment of a family and is part of a communal environment:
A home has an interior and an exterior:
Terminus of a series of communal services:
Start of a personal enterprise.

Fig 2.1 The Two Spheres of Housing, from *Three R's For Housing* by N.J. Habraken. Reprinted with permission.

tion of abutting dwellings. Habraken proposed to create such support physically, technically and organizationally. 'Supports' would provide access to common mechanical systems, accommodating a variety of dwelling unit plans. These three-dimensional structures would enable detachable dwelling units to be installed independently from – but supported by – the base building.

Habraken's approach contrasted sharply with contemporary proposals of the Japanese Metabolists, French architect Yona Friedman, the 'Townland Project' in America's Operation Breakthrough, the proposals of the group SITE and scores of related initiatives. It had, in fact, a precedent in a proposal outlined by J. Trapman, another Dutch architect whose 'kristalbouwen,'or 'tall constructions in oblong blocks' advocated fixed and demountable supporting structures and flexible plans (Trapman, 1957). Habraken was not interested, however, in reinventing housing types. He proposed to construct Supports in full recognition of and in harmony with local culture, including traditional architectural character. He unequivocally advocated the use of appropriate local technology, although questions of technology by no means governed his proposal. In addition, Habraken sought to directly connect with contemporary professional practice and, in so doing, to transform it. His vision was methodological as well as formal.

Habraken further advocated clearly separating domains of responsibility and decision-making authority between the community Support and the individual **detachable unit** (the latter comprising those technical systems and consumer items needed to fully inhabit an empty allotted space within a Support structure). He believed that the organic, fine-grained qualities of a viable neighborhood – precisely what was missing in mass housing – might then regenerate over time.

Mass housing had also failed to provide any appropriate mechanism for industrialization of the housing industry. Now Habraken suggested that the production of Supports and detachables could finally harness the efficiency of industrial manufacturing. To meet the huge and differentiated demands of individual households, a market for diverse consumer 'infill' dwelling products distinct from the Support would develop.

2.1.2 Theory, methods and implementation

Within the space of several years, Habraken's theories and proposals had begun to be disseminated worldwide, via word of mouth, articles, and unauthorized 'bootleg' translations of his seminal Dutch-language book. A decade passed before Habraken's *Supports* was commercially published in English. By then, the Supports movement was growing international in scope and stature. Its growth was chronicled in *Plan, L'Architecture d'Aujourd'hui, AD, Architecture Review, Toshi-Jutaku* and *Open House* (now *Open House International*). At the center of the Supports housing movement was the Stichting Architecten Research, or SAR (Foundation for Architects' Research). SAR had been founded in 1965 by an alliance of Dutch architecture firms, with Habraken as Director. Supported by contractors, industrialists and others, its official goal was to 'stimulate industrialization in housing.' More generally, it sought to study issues surrounding the relationship between the architecture profession and the housing industry, and to chart concrete new directions for architects in housing design.[1]

Based in Eindhoven, where Habraken had been asked to chair a new school of architecture, the SAR produced an extensive body of applied research, including *SAR 65* (basic methods for designing residential Supports without predetermining the size or layout of dwellings), and *SAR 73*, (a methodology for the design of urban tissues). SAR also introduced the 10/20cm 'tartan' band grid which eventually became a standard for modular coordination throughout Europe.

During the first decade of the SAR, it served as the incubator of Supports ideas and applied research, which were then disseminated throughout the world. Realized OB-like projects were occurring elsewhere, primarily in Germany, Sweden, Switzerland and Austria. All of that changed, however, in the mid-'70s. As realization of projects elsewhere dropped off, Dutch Supports began to be constructed in increasing numbers. The first had been Van Wijk and Gelderblom's 1969 housing complex in Hoorn. Beginning with a project of six experimental houses in Deventer, Age van Randen and the firm of Van Tijen, Boom, Posno, Van Randen realized another half dozen Support projects over the next four years. Fokke de Jong, Hans van Olphen and Thijs Bax of J.O.B. Architects were building projects and proposing urban tissues based on SAR theory. And Frans van der Werf/Werkgroep KOKON had begun to construct the first dwellings in Van der Werf's quarter of a century of Open Building work.

After a decade of applied research and publications followed by pioneering prototype 'Supports' projects built by municipalities and housing associations, the groundwork for Open Building had been laid. There was substantial ongoing funding from both the Ministry of Housing, Spatial Planning and Environment and the Ministry of Economic Affairs. Additional research groups continued to contribute important work. These included IBBC-TNO Bouw, Vereniging van Systeembouwers, VGBouw (Vereniging Grootbedrijf Bouwnijverheid), the Faculties of Architecture at the Technical Universities of Eindhoven and Delft, SBR (Stichting Bouw Research), and KD Consultants BV.

Theory disseminated by the SAR had come to form part of the mainstream discourse of architecture, and the realized OB projects were celebrated. Lucien Kroll's Maison Médicale student housing at Louvain, Belgium, popularly known as 'La Mémé,' was celebrated throughout the world. Even decades later, the *Architects' Journal* would describe Hamdi and Wilkinson's second PSSHAK project at Adelaide Road in London as one of the most influential projects of the '70s. In the US, Frans van der Werf's article on the Papendrecht project in the Netherlands was featured prominently in the inaugural issue of the *Harvard Architectural Review.* There had been no realized Open Building projects in the US, but a generation of 'housers' had come of age with *Supports,* and theorists including Robert Gutman, Kenneth Frampton and Christopher Alexander were discussing 'Supports theory.'

At the beginning of the SAR's second decade, John Habraken departed for America, where he assumed the leadership of MIT's Department of Architecture. The SAR's direction and focus now shifted under the leadership of John Carp. Beyond the SAR's doors, things were also changing. New and critical factors came into play, not the least of which were the first short-term economic aftershocks of the oil shortage, cyclical shifts in the political climate, and profound permanent changes in the production of housing. The focus of Dutch architects in the Supports mainstream was also evolving: as they moved from discussing theory to building projects, new practice-oriented issues became preeminent.

Within the SAR, work continued on research projects and publications to add to the already substantial literature on Supports, urban tissues, design methods and related fields. But beyond the SAR's walls, the dearth of infill systems continued to mean that full implementation of Supports theory simply could not occur. While the elegant Supports theories and practical SAR methodological tools remained, the Foundation's base of support was narrowing. During the SAR's ascendancy, the leadership role and investment of both national governments and the architectural profession in housing had peaked; it was now rapidly diminishing. As it waned, so too did the SAR's influence in housing research and debate in the Netherlands. Increasingly, there were debates surrounding the fundamental practicality of building Supports. Some architects within the movement began to feel that Supports could not be implemented. Some felt the theory was becoming outmoded in light of fundamental shifts in the social organization of housing, the housing market, and the overall political climate.

During the period between 1974 and 1982, projects in the other nations dropped off. However, in the Netherlands, increasing numbers of projects were constructed. From 1970 to 1982, 20 residential OB projects were realized. Of the 16 projects built during the decade of the '70s, 13 had been realized by pioneers of the Open Building movement: Van Randen, De Jong, Van Olphen, Bax and Van der Werf/Werkgroep KOKON.

In all, over the course of two decades, 50 projects had been realized; 40 in the industrialized West, and 10 in Japan (the latter constructed between 1980 and 1984.) Subsequently, while Japan leapt forward into the production of Support/Infill housing, production of Supports projects dwindled throughout Europe. After a decade, Germany, Austria, Switzerland and Sweden had all but abandoned Support housing; it seemed in danger of becoming an isolated Dutch phenomenon. And at that, in the years between 1982 and 1989, there were only three Supports projects built in the Netherlands, all of them by Frans van der Werf/Werkgroep KOKON.

Among the core of highly experienced architects of realized Supports projects, there was unanimous agreement: If Support methods and principles were to succeed in broad-based practice, problems associated with technical and procedural entanglement remained to be solved. Pioneering architects, contractors, and clients had demonstrated their ability to overcome considerable obstacles, to actually build Support structures. Analyses by Karel Dekker and others had shown conclusively that Supports represented a viable economic model. But the builders of Supports found no economically viable fit-out systems to install within them – important groundwork research and product development by corporations including Bruynzeel and Nijhuis notwithstanding.

For many in the Netherlands, the new direction clearly became implementation. Like other forms of residential construction in the industrialized West, OB projects were hampered by post-war patterns of production, and by organizational, social, technical and conceptual inertia. There still existed no replicable mainstream residential precedent to serve as a model for organizing large-scale Open Building practice. Infill technologies developed for the commercial real estate market were clearly successful, but they were not yet compatible with the residential market.

Government ministries continued actively funding grants to stimulate innovation in the building and housing industries, and research followed suit. Supports investigations shifted toward infill production, regulatory reform, and the architectural technologies of Supports. As the oil crisis ebbed, the real estate development and residential consumer markets began to reinvigorate. Some 'Supporters' now forged active links with industrial manufacturers seeking to enter those recovering markets. Starting in 1975, the Van Randen Group had begun to actively conduct applied open architecture research at the Technical University of Delft. Now a new term and a new movement – Open Building – emerged. A partial inheritor of the SAR's legacy, the OBOM (Open Bouwen Ontwikkelings Model, or 'Open Building Simulation Model') Research Group at the University of Technology at Delft was formed in 1984 under Age van Randen.

By 1984, the Board of the SAR had shifted its focus not so much toward implementation, but abroad. Foreign interest in SAR methods and research might be ebbing in northern European nations, but it was flowing toward other parts of the world, including Latin American and Asia. The SAR still witnessed a constant stream of international visitors passing through its doors, or attending courses at the Bouwcentrum in Rotterdam. In 1985, SAR Director John Carp helped mount an initiative to launch Network: International Foundation For Local Housing and Design Research, to extend the base operations of the SAR outside the Netherlands. An initial series of international summits, and the deployment of a pilot implementation team to Mexico City, seemed to hold great promise. Nonetheless, a formal worldwide structure could not ultimately be sustained. Subsequently, work continued on other projects,[2] publications, educational programs, and applications until the SAR closed in 1987.

OBOM, in the meantime, had set out to move Open Building toward broader and more complete implementation. An important first step was to further clarify the separation of residential base building and fit-out. More hardware, more know-how and regulatory reform would be needed to develop and implement fully prefabricated residential infill technology. In the years which immediately followed OBOM's creation, comprehensive infill systems began to be developed in the Netherlands. OBOM has continued up to the present to investigate key questions related to Open Building implementation, and to disseminate OB information worldwide.

Prior to the SAR's creation, non-residential Open Building had begun to gain momentum, spontaneously and without ceremony. By the time the SAR closed its doors, commercial Open Building had become conventional practice. Commercial, retail and institutional built environment throughout the industrialized world was irrevocably transformed in the process. In that sense, Habraken's visionary prediction about the reemergence of a mutable infill level, a level which would restore the natural relation between user and building, and the historical balance between social constancy and individual freedom to transform, had come true. But it was not yet occurring in residential architecture.

2.2 JAPAN

2.2.1 Postwar introduction of industrialized housing

It is important to note that industrial capacity for building construction developed somewhat later in Japan than in western countries. For many centuries, indigenous building had been based on timber frame construction within a highly-evolved craft tradition. Construction also followed a shell/ infill or 'two step' principle: carpenters first raised the frame and closed it in, only later building the infill – tatami mats, sliding shoji screens, etc. - to separate spaces. Mid- and high-rise multifamily housing was quite rare.

Rapid urbanization and population displacement followed the Second World War. In its wake, multi-story multi-family housing became a dominant new form arising in the Japanese landscape. At the same time, ownership was joined by renting as a prevalent form of tenure. Multi-family housing had had a short history in Japan: widespread traditions of multi-family housing had developed only since the establishment of the Japan Housing Corporation (later to become the Housing and Urban Development corporation) in 1955.

With the introduction of new housing types from the West, bolstering the building industry's technical capacity to prefabricate and build in concrete, steel and glass become a nation-wide concern. In the 1950s and '60s, many leading architects and government officials, including Mitsufusa Sawada of the Ministry of Construction (MOC), and Professor Yositika Utida of The University of Tokyo, championed the simultaneous development of a strong open market of building materials and of industrially-produced 'standard' subsystems for housing and other building types. Jointly, they set in motion a number of projects which began to transform the entire construction industry. The accompanying changes to lifestyles, economics, land use, professional services, construction processes and technology were dramatic.

In succeeding decades, researchers in universities, government agencies and private corporations continued to study, experiment with and develop new housing components, systems, supply lines, and production methods. Both the new housing forms and the more traditional wood-framed housing technology used in detached houses were affected. Gradually, a number of distinct but interrelated strands of development took shape. These included:

- modifications to conventional wooden housing;
- development of manufactured housing;
- development of American-style wood stud framing;
- emergence of a renovation market; and
- development of technology for mid- and high-rise housing framed in steel, concrete and composite materials.

Within these interwoven strands, wood frame construction and commercialized manufactured housing were in direct competition. Initially, little information concerning technical systems and methods was divulged between them. Between other strands, however, important exchanges of technical methods were made, benefitting all housing production.

Methods for building residential structures in full compliance with the rigid requirements of seismic design and fire protection evolved rapidly. In addition, important developments took place in cladding systems, mechanical systems, interior components such as unit bathrooms, and fixtures and finishes. Development of many of these components was actively supported by the Center for Better Living (BL), an independent organization chartered by the Ministry of Construction, which would ultimately certify many of them.

Within the short span of 30 years, major developments had occurred in prefabrication techniques and industrial production of building elements.

All of this activity combined to elevate Japan to a relatively sophisticated level of industrial development with respect to housing. Within those several generations, Japanese built environment and its social processes were also utterly transformed. Such rapid change, unprecedented in the industrialized West, was the result of concerted action in both the private and public sectors. Throughout all the changes in methods and systems, the traditions of conventional wooden house construction remained a strong and influential undercurrent as other construction methods were introduced. Yet the number of traditional carpenters declined rapidly and with them, the number of traditional houses.

By 1972, the number of dwelling units built in one year reached nearly two million. This marked the pinnacle of Japan's mass housing era. That period was characterized by a focus on maximizing the production of units to meet fundamental and pressing housing needs. One long-term result was ultimately the creation of a substantial stock of small and often poorly built units, many in large housing blocks. In addition, the durability of housing built in the immediate post-WWII years emerged as a national problem in the 1980s. The repair and maintenance of sanitary equipment and bathrooms, and the upgrading or replacement of piping and wiring embedded in the floors and walls of condominium and apartment buildings had become a major hurdle. So was the buildings' social structure. Traditional neighborhood organizations such as the *cho nai kai* no longer bound neighbors on different floors. Multi-family housing was popularly viewed as a transitional stage in the housing cycle. Families somewhat universally hoped to raise children in freestanding single family residences.

The Japan Housing Corporation was instrumental in early efforts to reorganize and strengthen the participatory role of households in multi-family environments. After several decades, the cooperative housing movement began to develop in both Tokyo and Osaka, with the aim of supporting individual households in meeting their own housing needs in multi-family housing. In 1978, the Japan Coop Housing Association was formed. Subsequently, in the 1980s, a number of cooperative projects were built in Osaka under the name ToJuSo.

Other principle interests in Japan included the development of new components, production methods, and supply methods for housing. Much effort and investment began to focus on detached housing. Additional efforts focussed on components for multi-family buildings, including new structural systems, cladding elements and interior components, and planning and design methods that would take into account the interests of occupants.

At the same time, as a result of the very high price of land and housing in big cities in Japan, homeowners or condominium owners increasingly found it difficult to move from one house to another. Couples who hoped to raise families commonly aspired to move to detached houses as soon as possible, but found their aspirations thwarted. Now, households were frequently constrained to live for many years in very small postwar units with generic floor plans. The limited housing stock and escalating prices placed a particular hardship on the growing number of elderly households.

In setting the stage for a brief survey of the multiple interwoven strands of Open Building in Japan, it is also important to briefly compare Japan's national building culture with that of the Netherlands. Both had dense low-rise patterns of housing. Both building traditions had evolved slowly over hundreds of years, one employing timber frame and the other a combination of masonry and timber. Both had longstanding and sophisticated housing guild traditions. Both were highly organized and modularized: Japanese construction was coordinated by the *ken,* roughly about 1.8m. Dutch construction expressed door and window measurements in brick modules. Builders in both nations constructed typologies which were known to all: the proportions of a six-tatami room, or of a *voorkamer* in Amsterdam were well established.

One key difference between the cultures is that Japan has no history of individualism or market independence parallel to the Dutch experience. Nor is there a clear separation of government and private sector. Operating at an economic scale many times that of the Netherlands, Japan's extremely powerful Ministry of International Trade and Industry (MITI) and its Ministry of Construction (MOC) actively originate, support, shape and manage large scale initiatives across multiple industries. This is frequently accomplished through so-called Third Sector (public-private hybrid) organizations. Confluence of interests frequently leads interested parties to participate in projects with several overlapping roles. By comparison, in the Netherlands, private foundations such as the SAR and OBOM seek and respond to opportunities for government funding, and the organization and research and development of infill companies has been occurring at a small scale. OB clients in the Netherlands until very recently tended to be small scale local housing associations, local developers and small builders.

In Japan, many interlocking or complementary initiatives are directed toward the entire housing industry. They are fundamentally government initiatives, undertaken with the coordinated participation and sponsorship of Japan's largest builders, housing and development corporations, component manufacturers and associations. MITI has tended to focus on the infill level. Toward that end, it has created and/or invested in developing infill component industries. MITI's efforts have produced leading research in energy conservation systems, demountable partitions and storage systems and other interior components. The MOC has tended to focus and organize research and funding on the level of the base building, infrastructure and building codes.

2.2.2 First interest in Open Building

The earliest Japanese 'missions' to the SAR, and the earliest broad dissemination of western Supports theory in Japan, occurred in the 1970s. Early in the decade, Japanese research architects visited the SAR in the Netherlands. Widespread interest in western Open Building began to surface, led by Yujiro Kaneko of the Building Center of Japan and Seiji Sawada. In the years that followed, further visits were made to the SAR and its successor organizations, including OBOM. In September 1972 and again in January 1979, *Toshi-Jutaku* published detailed reports on Support/Infill concepts, architectural methods and realized European projects. Other 'systems building' experiments in the United States and Europe, including CLASP (UK), SEF (Canada), SCSD (US) and others, were also the subject of close scrutiny during these years.

A number of important experimental projects for school buildings (e.g. GSK) and other building types (e.g. the GOD project) were constructed as demonstration projects to stimulate and develop 'systems building' in the Japanese market. In 1980, a first Open Building project was realized in Japan. Three more followed in 1982. Subsequently, more than three dozen OB projects, ranging in size from a handful of experimental dwellings to dozens of multi-family buildings, have been realized in Japan. Many more incorporate some of the specific goals and technologies of Open Building.

As is explored in Parts Three and Four, some fundamental differences in approach to OB characterize developments in Europe and Asia. Within Asia, there are also significant differences between Chinese and Japanese approaches to Open Building. Although developments toward Open Building in western Europe and Scandinavia have been widespread and diverse, Japan as a nation is unique in the diversity of its approaches, as well as in the scale of its investigations of Open Building, the number of distinct institutional long-term players involved in parallel developments, and the unprecedented number of projects and dwellings produced.

2.2.3 HUDc/Kodan Experimental Project (KEP)

Of the major pioneers of OB in Japan, one of the earliest and most persistent contributors has been the Japan Housing and Urban Development corporation (HUDc). To further the development of systems approaches to housing, HUDc implemented a three-phase, six-year project called the Kodan Experimental Project (KEP) in 1974. Building on a Japanese traditional architecture that is inherently open, KEP built what were perhaps the earliest realized professional project developments toward Open Building in Japan. As a secondary objective, KEP's initiators decided that in addition to fostering experiments in hardware, a new generation of researchers with new approaches would be needed for the future.

The KEP project team divided the building into five subsystems: structure, skin, interior finishes, service or sanitary systems and air conditioning equipment. For each subsystem, performance specifications were prepared and component manufacturers organized to develop suitable new components. A 300mm grid and interface rules were used to enhance component interchangeability. The resulting product catalog was prepared as a design tool.

The first experimental structure was erected at the HUDc research laboratory in Hachioji, west of Tokyo. Many projects and initiatives have followed, focusing on lowering costs, rationalizing construction and increasing consumer choice in housing. Major realized building programs have included the Free Space Support System (1983); the Free Plan Rental Project (1985); the Green Village Utsugidai group condominium project (1993); 'You-Make' Cost Reduction Model Housing (1995-7)[3]; and the KSI Experimental Project (1998-9). Many of these milestone projects are explored in case studies in Part Two.

2.2.4 Two Step Housing Supply System and Century Housing System (CHS)

Another important development toward Open Building in Japan has been the implementation of Two Step housing delivery. Kazuo Tatsumi and Mitsuo Takada (Kyoto University) first studied existing urban housing. Their research demonstrated that the absolute boundaries drawn between 'public' and 'private,' and between 'owned' and 'rented,' were neither sufficient nor accurate when applied to urban housing in Japan. A new distinction was needed, one which correlated with the complex characteristics of housing products and processes. The researchers' collaborative work led to the development of the Two Step Housing Supply System, which builds on

Japan's rich vernacular tradition of shell/infill construction. As a first step, a public Support is designed as **social overhead capital**, common property of good quality and long durability. In the second step, the infill is installed, ideally supplied by small regional construction companies.[4] The driving idea is not so much rationalizing construction as it is more effectively organizing the social sphere of housing.

The first Two Step project was built in 1982 in Senboku New Town, Osaka. In this project, the positions of bathroom and kitchen were fixed, but other parts were freely arranged for each occupant. Planning sheets were prepared for user participation. In 1989, the Senri Inokodani Housing Estate Two Step Housing Project, was built (also in Osaka). Senri Inokodani employs a utility trench in the middle zone of each unit to locate 'wet' cells with some variation. The rest of the unit can also be arranged in a variety of ways. This project also utilized a variation of the Century Housing System (CHS), developed by an MOC committee chaired by Yositika Utida. More recently, the Hyogo CHS and Yoshida/Next Generation Urban Housing Projects combined the Two Step and CHS concepts. The structural systems use an inverted slab/beam floor system similar to that in several other projects of that period. (ed Kendall, 1987)

The CHS started in 1980 as a five-year initiative supported by the Ministry of Construction to extend the life of new housing stock, both physically and functionally. The project committee led by Professor Utida adopted a building component system based on modular coordination and the concept of **durable years** related to each component group. The idea of modular coordination was already well established in the building components market for conventional wooden houses. The CHS team aimed to make coordination rules that would apply to both conventional wooden construction and the more recent construction in reinforced concrete.

In the CHS approach, the first principle regards the interface between two different component groups, and their construction sequencing: components whose durable years are relatively short must be installed only after components with longer durable years. For example, piping and wiring must not be buried in cast concrete or other structural components. This led to the use of raised floors.

Research was carried out from 1980-82. In 1984, the HUDc developed several five-story buildings for foreign residents who worked for the International Science Exhibition held in Tsukuba, north of Tokyo, in 1985. All 159 units were later completely refitted with infill to accommodate more typically Japanese life styles. In 1985, a condominium project in Tokyo utilized trenches to allow for horizontal piping distribution within units. Construction of a CHS condominium of 263 units in Tokyo in 1986 was followed by another condominium in Tokyo in 1987. In 1989, the Senri Inokodani Housing Estate, also using the Two Step Housing Supply System, was built in Osaka.

In 1997, yet another project combining Two Step and CHS was built in Hyogo Prefecture near Osaka. In the latter project, the slab and beam assembly was inverted, leaving a flat ceiling in the unit below and upturned beams below a raised floor in the dwelling unit above, allowing easy routing of piping and wiring within that unit. Dozens of CHS projects have been built, including detached houses and multi-unit apartment and condominium buildings. The most prominent Open Building project to date in Japan is the Next21 (1994) project sponsored by Osaka Gas Company. It is a combination of Two Step Supply System and the Century Housing System, and includes, as well as advanced building systems, a number of important experiments in sustainability, including energy conservation, recycling and establishment of urban green zones. It also represents the first OB project in which one architect designed the Support and many other architects designed individual units.

2.2.5 Tsukuba Method

In response to a number of problems related to 'right of use' laws concerned with land development, a series of projects based on the 'Tsukuba Method' began construction in 1995. This work was led by Hideki Kobayashi at the Building Research Institute of the Ministry of Construction in Tsukuba. The goal of this initiative is to implement a new concept of land ownership and household control, while using the Two Step Housing Supply approach. The first project accommodates 15 households, a second houses four and a third incorporates 11 units. Six more projects have been built by private companies, and more are planned, thus moving the concept out of the realm of government sponsorship into the larger – and most important – private market. By tackling the problem of land availability and cost, and by further linking that to the issues of living in one place instead of the normal cycle of moving, the Tsukuba Method has made a significant breakthrough toward OB in Japan.

3

A brief interpretive history of Open Building

3.1 FROM VERNACULAR PRECEDENTS TO OPEN BUILDING

As the previous chapter has illustrated, Open Building has developed in part out of reinterpretation of vernacular building traditions. OB's rationalized processes and strategies for delimiting boundaries of control extend traditions that are probably as old as the built environment itself. Similarly, constructing and enclosing structure against the elements prior to fitting-out the interior, and separating infill from base structure, makes practical sense, particularly in harsh climates. Because most vernacular building types experience a wide range of uses in their life span, builders learned long ago to make the infill level distinct, changeable, less enduring, wherever doing so did not compromise structural performance or the basic social understanding embodied in the building type.

Thus, traditional Japanese building used demountable and sliding screens and removable tatami floors between structural posts. Traditional Dutch canal houses first built the facade, roof and fenestration, then arranged rooms behind the windows. Warehouses anywhere might be constructed with perimeter masonry walls supporting wide-span wooden roof structures, with a largely independent timber frame infill, just as some Greek vernacular houses of masonry might have two-story timber infill mezzanine structures with wooden screens. In every case, each inhabitant in turn would interpret and exploit the capacity of the type (Habraken, 1998).

On the urban scale, freestanding buildings forming a continuous street wall traditionally maintain independent walls perpendicular to the facade: sometimes to transfer roof loads, but just as importantly so that each building can be demolished and replaced without unduly disturbing either abutters or the urban fabric. In cultures that build adjacent row houses on common bearing walls, that pattern incorporates elaborate legal and social agreements to define responsibility and control regarding the collective structural element.

For technical, conceptual, social and organizational reasons, conditions in high-rise residential buildings magnify the complexity of distribution, access, control, reverberation and responsibility. In conventional construction, such complexity generally precludes the granting of substantial autonomy to the individual dwelling, either in the interior layout or on the exterior facade.

As previously noted, recent centuries have witnessed unprecedented environmental change, including the widespread introduction of utility supply, waste removal, power and data transmission and transportation networks on all levels. These have been accompanied by profound changes in the scale, organization, purpose and regulation of environmental intervention. Buildings have opened up to diverse systems during a long-term shift from permanently installed hand-crafted parts to variable or easily changeable industrially-produced parts. This change has been attended by a consequent downward migration of many parts from the building level to the infill level. Ownership, control and responsibility are shifting from base building owner to tenant. The introduction of wave upon wave of building systems into buildings in rapid succession has resulted in a great deal of on-site improvisation and confusion. Such waves of environmental change explain the subsequent era of entanglement, the background against which present developments toward Open Building occur.

3.2 FROM MASS HOUSING TO OPEN BUILDING

Governments have designed and executed large urban interventions throughout history – including colonial cities, fortifications, public utilities, and transportation systems – and regulated much more. Such traditions are millennia old. But beginning with the Dutch Housing Law of 1901 (followed rapidly in other European countries, and in Japan and China after WWII), governments began to appropriate responsibility for protecting the health, safety, welfare and property values of dwellers, in response to two forces:

First, a new social attitude emerged with regard to public responsibility for individual and social well-being, enabled in part by changes in the distribution of increasing wealth. Second, industrialization, urbanization and the introduction of new, complex and potentially hazardous environmental technologies, infrastructures and networks, coupled with increased awareness of dangers, required public oversight of previously private activity.[5] In many European, North American and Asian countries, governments began to define, regulate and enforce environmental standards. At the same time, many governments began producing mass housing. The phenomenon of mass housing brought together large numbers of standardized housing units; new high density typologies; new networks of supply, waste removal and transportation; and further centralized government decision-making and control.

Championed by government and institutional bureaucracies, mass housing spread throughout capitalist and socialist societies alike during the era between and following the 20th century's world wars. The urban house lot, the basic building block of urban fabric, was replaced by coarse-grained, multi-story housing blocks, often containing hundreds of rigidly uniform dwellings. Mass housing's 'top-down' professional intervention brooked no inhabitant participation. Its characteristic uniformity resulted from the application of a series of extraordinary and concurrent developments in other fields: Emergency housing production methods pioneered by the military in wartime were put to civilian use. Scientific management and production techniques, including Taylorist industrial organization and assembly line production, were adapted to the prefabrication of building components. Simultaneously, centralized decision-making was employed on a huge scale by new or expanded bureaucracies with unprecedented powers of environmental control.

In the contemporary cultural climate, applying rational scientific engineering seemed an obvious approach to 'solving' a perceived 'housing crisis.' Achieving large-scale efficiency through value engineering while improving hygiene, housing standards and construction efficiency became a singular goal for the institutions and professions driving mass housing. Generic building layouts and facades and inflexible unit plans were accepted as necessary by-products of rationalized and prefabricated production. In reality, they resulted from shifts in decision-making and control, particularly from decentralized patterns of responsibility to unitary, simultaneous centralized administration on many environmental levels.

By the late 1950s, mass housing sites worldwide, some barely a decade old, began to witness socially destructive effects from such dramatic coarsening of the urban fabric, centralization of control and attendant loss of individual freedom, participation and responsibility in built environment. In subsequent years, as rates of technical and social change have continued to accelerate, mass housing has proved inflexible, incapable of adjusting to social, economic and technical changes. Increasing numbers of precast concrete mass housing projects have now become obsolete or uninhabitable. Massive economic, social and environmental consequences have resulted. Fueled by such rapid unplanned obsolescence; issues of sustainability and choice; industrialization of housing subsystems; shifts from new construction to renovation and revaluing; downward migration of building subsystems to a reemerging infill level; and the emergence of a residential consumer market and of residential infill systems, Open Building has in recent years expanded well beyond its original constituency.

Over the course of a third of a century, over 130 separate residential initiatives, some involving dozens of multi-family buildings, have been constructed on principles of Support and infill. Countless other projects are beginning to incorporate at least some of the objectives of OB, in terms of tenant participation, consumer choice, flexibility for subsystem changeout, disentangling systems and decision-making processes by level and utilizing proprietary infill technologies.

3.3 KEY OPEN BUILDING CONCEPTS

3.3.1 Levels

As a result of almost four decades of investigation, there exists a substantial body of knowledge, theory and applied research related to environmental and decision-making **levels**. Behind it all is John Habraken's early, instinctive and relatively straightforward realization: the physical elements that make built environment are always directly associated with the actions of people: the two are inextricably bound; to consider dwellings as isolated objects or products leads to unacceptable consequences. That understanding ultimately led Habraken to another: as built form transforms over time, the shape of change reveals patterns of control. Whether we uncover abandoned plaster ceilings, filled-in doors and windows, or ancient stone walls, the patterns of change delineate hierarchical realms of control.

To create satisfactory and responsive long-term residential architecture, Habraken pointed out the need to understand who controls the form. The use of levels enables environmental professionals to define the environmental agents in control – who controls what, and when – as a fundamental criterion in designing. The levels schema thus allows distinctions to be made concerning the locus of control between individuals, group or organizations. The theory of levels also takes into account the fact that the particular parties exercising control often change between the phases of design, construction, occupation and the life of the building.

Although levels may shift and occasionally absent themselves from time to time and place to place, their existence is constant throughout built environment: levels are universal. Levels define themselves where certain groupings of physical parts and spaces can be observed to jointly transform in an orderly and recurring way. In essence, they form spontaneously at points where boundaries of construction, social organization and territory coincide. By analogy, they act like control joints: levels represent places where self-organizing and continuous built environment allows for breaks in the formal structure or in the control structure, precisely where such discontinuities will not disrupt the whole.

The abstract idea of an ordered hierarchical structure of environmental levels reflects everyday experience. Groupings within the hierarchy of levels are implicitly familiar to everyday inhabitants as well as to professionals. It is universally understood in the West that individuals can't build houses out onto the street, expecting the road to shift elsewhere. As a rule, furniture is purchased or planned to fit in specific rooms, in preference to constructing or shifting partition walls until one's furniture is well surrounded. Levels define both the environmental professions and their fields of operation – urban planning (tissue), architecture (base building), interior design (infill), furnishings.

Open Building's use of explicit notation of levels as a basis for steering environmental interventions in methodical ways represents a fundamental change in professional practice.[6]

3.3.2 Supports

In the most basic sense of the Support concept, just as a highway is a finished product whose lanes are intended to be occupied by many kinds and sizes of vehicles, a Support is a finished building, ready to be occupied by variable infill. However, the layout and size of individual occupancies – dwellings, offices, etc. – are not pre-determined. The Support is the permanent, shared part of a building which provides serviced space for occupancy. In terms of real estate and property ownership, the multi-family base building functions as vertical real estate, to be developed and subdivided as might any other real estate development. It will include public ways (stairs, elevators and corridors or galleries), commons (laundry rooms, community rooms, public foyers, etc.). **Parcellation** or subdivision will result in the creation of allotments or 'lots,' and services will be run to each of these directly from the public space, just as public utilities may service each rowhouse from trunk lines buried beneath the street or sidewak. Supports can be constructed in any durable materials, incorporating any technical systems. In all cases, they provide capacity to satisfy diverse and changing demands throughout their useful life. Supports are either newly constructed or made from existing building stock.

Supports contain all shared (common) building services, delivering them to the front door or party wall of each occupant. Typical support elements include building structure and facade, entrances, staircases, corridors, elevators and trunk (main) lines for electricity, communications, water, gas, and drainage. By contrast, dwelling unit heating and air conditioning equipment is not generally part of the Support. This avoids technical or social entanglement, the passing of public infrastructure through private space or the creation of unintended and undesirable conditions of hierarchical control.

The Support is dominated by the local market, architectural styles, climate, building codes and land use rules, investment requirements and other local conditions. Thus, within its specific social and technical setting, the Support is built using locally appropriate means of design and construction.

Transformations in building use result from social, technological, demographic and market changes during the useful life of a building. Even where residential use remains, initial unit sizes and configurations become outmoded as incomes, household composition and space needs change on a dwelling-by-dwelling basis. The Support is intended to accommodate and outlast infill changes, to persist largely independent of the individual occupants' choices, while accommodating changing life circumstances. It embodies values and preferences determined in common or by the initial developer.

A Support is not a mere skeleton. It is not neutral, but is rather enabling architecture. It is more like a serviced, environmentally protected site within a built landscape: a Support is a physical setting that offers space and possibility to make dwellings with as few constraints as possible, while requiring as little work as possible. A serviced plot of ground with regulated use, building placement and size restrictions is, in fact, a limited sort of Support. Furthermore, a new row house development in which a developer invites buyers to customize dwelling interiors within constraints is also a kind of Support, as are converted warehouses, schools or office buildings in which each unit is individually determined and later sold and altered again to a residential occupancy.

Once erected, the form of the Support is closed in. The common services are all installed, the site is cleared, and any disruption to local traffic is at an end. From the perspective of the community, the Support appears complete. But to be occupied, it requires infill.

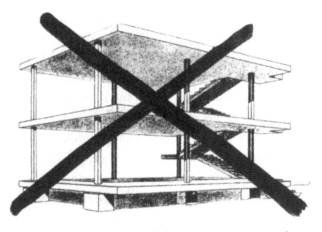

Fig 3.1 A Support is not a Skeleton. Image courtesy of N.J. Habraken.

3.3.3 Infill

The most important story of architecture in the last half of the 20th century, from a technical and organizational perspective, may be the evolution of infill. Infill has liberated architecture from problems of piping, wiring and ducting. In Open Building, most of the technical and organizational problems of these building elements shift to a lower infill level, with powerful effect. With the adoption of an infill approach, the roles of architects and consultants are altered significantly. Their work becomes more focussed on architecture, which may be defined as the durable common part of buildings. By definition, each dwelling's infill is independent. Whether purchased or leased, it is under the control of the resident.

Infill systems already exist throughout the commercial office market. Several independent companies – such as the US-based Steelcase – and consortia – including Haworth, Herman Miller, Interface, Tate, Armstrong and other major furniture and interior product manufacturers – now market a range of products, from compatible components to 'slab-to-slab' fit-out systems. In some cases, they provide design services as well. A residential infill system is similar in concept to a commercial office fit-out, but more complex. It is more densely packed with mechanical and other supply systems. As a consumer product, it must meet the demands of a wide variety of individuals in an equally wide variety of base building types.

It is quite possible to fit-out a residential space in a Support using conventional construction. Infill elements need not be industrially produced. At present, both new construction and revalued building stock is fitted out with conventional residential infill, without systematic organization. From an organizational perspective, site-made partition walls are infill elements if the resident has control over their position, or if they can be changed independently of the Support without impacting any other dwelling. However, if the dwelling lease prohibits moving any element, it remains part of the Support, despite the technical ease of moving it. Thus, infill elements are defined by social as well as technical criteria.

An **infill system** is not an assortment of products brought separately to the site, each cut and installed by its own subcontractor. Rather, it is a carefully pre-packaged, integrated set of products, custom prefabricated off-site for a given dwelling and installed as a whole. Comprehensive infill systems provide the partitions, mechanical installations and equipment, doors, fixtures, cabinets, finishes and other elements needed to make a completely habitable space within a base building. Although infill systems and their components need not be industrially produced, separating infill from the base building encourages the adoption of sophisticated industrial processes, including interfaces, logistics, quality control and information management from other industrialized, consumer-oriented sectors.

Presently, OB projects around the world tend to have partial infill systems provided by a variety of companies and trades. Although the work is organized on a unit-by-unit basis, the logistics process is otherwise somewhat conventional. After the infill is selected, parts for each dwelling are manufactured, assembled, or acquired from various companies. Interfaces between parts are usually coordinated beforehand, thus leaving minimal cutting and fitting to be done on-site. Parts arrive on site independently, to be installed by independent trades coordinated by a conventional general contractor.

3.3.4 Unbundling decision-making

Presently, in most conventional projects, dwelling units are completed as part of a single building contract. Products are ordered for the whole building and installed throughout each floor, trade-by-trade. In North American wood frame construction, this process is often characterized by utter disarray. Such entanglement – accompanied by the confusion which has been a part of housing production since utility systems began to migrate indoors – does not lend itself to advances in systematic design and industrialized production. Nonetheless – and despite projects' increasing size, numbers of

Fig 3.2 Entangled building systems. Photograph by Stephen Kendall.

physical parts and numbers of decision-makers – architects and other professionals throughout this century have frequently ignored environmental
and building trends, continuing to advocate integration of the many separate decisions into one 'bundle'.

Separating buildings into the distinct bundles of technology and logistics – Support and infill – and the two domains of production related to
each, organizes production capacity effectively, developing each 'technology
bundle' or level's possibilities for optimal production. That in turn encourages systematic product development for broad and varied markets, a basic
prerequisite for industrial production.

Success stories of total building integration accompany either relatively modest projects or highly centralized control. Achieving integration for
an entire building is frequently a nightmare of socially and technically
unwieldy complexity, marked by conflict from the beginning of the planning process, through construction, facility management and beyond. As a
direct result of the confusion and rigid entanglement which results, future
adaptation becomes limited and difficult.

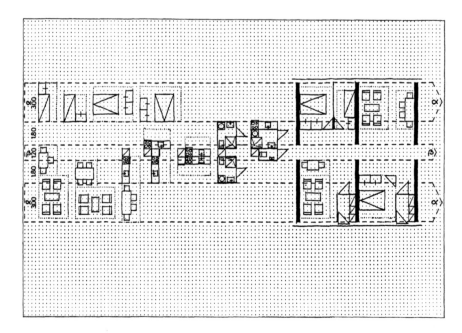

Fig. 3.3 Keyenburg capacity study drawing. Courtesy of Frans van der Werf.

3.3.5 Capacity

Conventional wisdom holds that designing starts with 'defining the problem.' 'Getting the program right,' then leads to a 'design solution.' In Open Building practice, **capacity** replaces the set program and its functional specificity during initial design. Capacity analysis is a complex and demanding practice at the core of Open Building. It is founded on two ideas: 1) designing form to be an open-ended and dynamic fabric; and 2) designing space or form (at multiple scales) with built-in capacity to accommodate more than one 'program of functions' over time. Open Building methods suggest that evaluation of an image shared by designer and client is the first step, one which inherently contains the germ of many alternate programs. Form is considered in terms of possibilities rather than in terms of a single, rigid and predetermined function. This in turn reinforces the concept of levels: a form (e.g. base building) may be judged based on its demonstrated capacity to accommodate multiple arrangements of lower level forms (e.g. alternate uses and interior layouts). Rooms exhibit capacity to allow multiple furniture arrangements and activities, and urban tissues may maintain coherence while accommodating a variety of building types and styles.

Design of the Support ideally incorporates capacity according to three principles: First, each dwelling in a Support must allow a variety of layouts. Second, it must be possible to alter the floor area by changing the boundaries of units within the base building or expanding it. Third, the Support or its parts must be adaptable to varying functions, some of which may be non-residential in character. Which criteria any given Support will have to accommodate becomes a function of project economics, site conditions, the preferences of various stakeholders and so on. The relationship between possible uses and their cost can be fully evaluated once the basic layout variations that a Support can hold have been documented.

In designing Supports, such evaluation of capacity must be approached systematically. First comes the evaluation of possible uses. This is a complicated process involving the comparison of a series of layouts. It normally begins with schematic design and follows throughout technical design. The interplay between base building and infill must also be explored. Since adaptability is an essential characteristic of Supports, change must be easily effected. Supports must be designed without knowing which particular infill products or systems will be employed, just as infill systems must be developed without knowing where they will be installed. Nonetheless, the form need not be neutral to optimize useful capacity. Totally 'flexible' multi-purpose space – space devoid of columns, walls, changes in section or qualities of light – offers no architectural definition for dwelling.[7]

3.3.6 Sustainability

At the height of the mass housing era, buildings throughout Japan and the West were for several decades rapidly constructed to minimal technical, structural and space standards. Such dwellings within 'unibody' multi-family buildings proved utterly incapable of adapting to subsequent lifestyle and technological change. Several decades later, remnants of the 'housing crisis,' the 'scrap and build' mentality and the short-term investment strategy behind them, are now widely rejected in favor of a 'building stock' approach. There is now more emphasis on basic principles of sustainable development, on maintenance and on refurbishment, if not yet on building for change.

A substantial body of research on developing re-usable components has further linked Open Building and sustainability. Building-in additional value in the form of long-term capacity for change poses a viable alternative to investment incentives and valuations based on short-term value extraction from real estate. Utilizing variable infill products forestalls rapid obsolescence of the entire building. In distinguishing between the parts of a building that should endure 100 years and those parts that realistically cannot have such a long life, Open Building creates a physical and procedural distinction. As one result, it is now possible to provide accurate life-cycle accounting of value and responsibilities commensurate with principles of sustainable design.

A further alignment between Open Building and sustainability concerns the development of technical interfaces that allow the builder or end user to 'plug-and-play' with products made by different companies. Manufacturers of architectural systems for the office market have led the way in bringing to market numerous products with standardized interfaces. But even these products are not recombined with products from other manufacturers, nor reused in new circumstances. Nonetheless, as products with high degrees of interconnectivity within specific product lines, their re-use value is heightened. By contrast, conventional piecemeal fit-out in residential or office buildings, while sustaining the base building, must be discarded with each subsequent reconfiguration. Such products have little engineered capacity for reuse, and, lacking any alternative, are frequently destined to add to construction waste landfill. Therefore, Open Building infill is moving toward design and manufacture for assembly and disassembly, supported by basic strategies such as 'click-together' components.

Technical issues aside, there remain questions of sustainability relating to social and individual choice and values. There is good reason to believe that the increasing physical entanglement of complex built environment constrains attempts to balance group and individual domains and responsibility. Sustainability concerns in great measure what is held in common. It represents community values, interests, and power to act on the community's own behalf. When it is not clear who is responsible for which parts of the physical fabric, any accounting for common purposes is almost impossible. When the commons is indistinguishable from the individual's territory, the means available to sustain what is shared are also few. Entanglement therefore thwarts advances in the evolution of sustainable built environment.

3.4 DEFINING AN OPEN ARCHITECTURE

3.4.1 What various professions gain from Open Building

Professionals adopt OB on different environmental levels in many nations and practice settings. Diverse concepts, products, methods, and best practices linked to Open Building are emerging worldwide in interrelated disciplines as OB is employed with different participants, methods, processes, emphases and outcomes. Following is an outline by profession of some of the reasons why professionals embrace OB strategies.

3.4.1(a) Urban Design

Open Building methodologies at the urban level include specific calculation techniques. Like the New Urbanism, OB also combines specific graphic notation methods with written performance requirements to ensure clear communication among implementing parties. In large projects, where distributed control is the norm, Open Building methods help to resolve coordination difficulties.

3.4.1(b) Architecture

Open Building offers a viable alternative to the prevailing conventional practice of adopting a single program based on unsubstantiated projections through time, wrapping the result in built form and then knitting mechanical and structural systems into and around the functions. Open Building also offers methods for incorporating decisions by all stakeholders, on all levels, throughout design and construction. By clearly disengaging decisions regarding internal spatial organization and utility services from the building's serviced shell, OB methods help to reduce friction among members of the project team throughout design, production and long-term management.

3.4.1(c) Interior design

Distinguishing between 'base building' and 'fit-out' assigns a specific scope to the interior design profession and the related product and installation industries. It establishes a new cluster of products and defines new kinds of production activity.

3.4.1(d) Product design and production

Developments toward Open Building have shifted increasing amounts of customized finished or ready-to-assemble (RTA) production to off-site locations, including prefabrication yards (where made-for-use elements are produced) and to manufacturers' facilities (where made-for-sale elements are made). Product design and manufacture are rendered more effective in supporting on-site work by observing a practical distinction among kinds of products according to level.

3.4.1(e) Contracting and construction management

In OB projects, much of the complexity of mechanical and utility systems shifts downward from base building toward infill. As one result, the efficiency of both base building construction and infill installation work increases. Base buildings can be simple and repetitive, built with speed and quality control, while maintaining capacity for great variety. The reduction of coordination among trades significantly reduces management overhead costs.

3.4.1(f) Finance and development

Open Building enables strategic combination of on- and off-site production, while controlling life cycle costs over the life of the building. As finance and development begin to more accurately monitor and evaluate investments by accounting for variable time factors, OB accounting methods make the added complexity manageable.

3.4.1(g) Public housing agencies

Open Building principles help housing agencies build projects that respect the budgets and preferences of individual households. When this is done, the economic power of each household is applied directly to the individual dwelling. The cumulative effect is magnified and reflected throughout the community, helping the social asset (base building) to remain healthy.

3.4.1(h) Facilities management

In typical apartment buildings, systems are largely entangled: changing one rental unit inevitably disrupts others. In the most legally burdened form of housing occupancy, the condominium, prevailing design and construction practices make every upgrade or changeout complex. Open Building principles eliminate or greatly reduce the number of conflict-prone limited common elements – greatly simplifying facilities management.

3.4.1(i) Sustainability proponents

Applied to the design and construction of buildings and neighborhoods, the fundamental principle of sustainability – considering the consequences of today's actions on tomorrow – leads to two open architecture imperatives. First: Build environments that can change, and thereby remain viable. Second: Disentangle subsystems so that the change or removal of one doesn't require the destruction of another – at the very least, design and build to reduce collateral damage. Open Building's key precepts and methods are aligned exactly with these basic directives of sustainability.

3.4.2 Common characteristics that define Open Building

A broad range of players with quite varied aims adopt Open Building practices to achieve different ends. Therefore, no strict measure of 'open-endedness' (Rappaport) is universally adopted. Nonetheless, residential OB does imply some basic conditions, such as a heightened degree of autonomy for the individual dwelling, its layout and equipment. Households in OB projects frequently exercise control when creating or changing their dwelling floor plans, and perhaps their units' facades. Or the building owner may exercise control by adjusting certain units to meet changing market conditions, without disturbing other units. In all cases, the individual unit remains physically distinct from the common property.

There is no clear dividing line in the continuum between open and traditional residential architecture, nor is any project fully 'open' in all ways. Rather, every realized Open Building project presents developments toward Open Building. What, then, defines open architecture? Or, more specifically, what defines a residential structure as an Open Building project? The answers are not constant from one era or building culture to another. Even within groups that practice Open Building, there is not always consensus. Nonetheless, over time, the professional practice of open architecture has come to be defined in relation to a finite catalog of specific approaches. (Tiuri, 1997; Beisi, 1998) These include:

1 *Recognizing and organizing work according to environmental levels*
2 *Distributing decision-making*
3 *Physically separating support, infill and other environmental levels*
4 *Disentangling building subsystems*
5 *Structuring professional services in support of household choice*
6 *Using specific Open Building methodological tools*
7 *Using specific Support technologies in conjunction with infill systems*
8 *Using specific infill technologies*
9 *Using specific Open Building financial instruments*

	VVO/Laivalahdenkaari 18 –95	Villa Paavola –95	Laivahdenportti 3 –96	Tammistonpiika –96	Lounaispuisto –96	Myllypelto –97	Meritähti –97	Laivalahdenkaari 9 –97	Linnanrakentajanpuisto –98	Rastipuisto	Tervasviita
User as decision maker											
A1 User decides on floor plan with infill	O										
A2 User participation at the support level	O										
B1 Optional floor plans for the first user			●	●				●		●	●
B2 User participation without changeability		O					●				
Open spatial structure											
A3 Regulation of the distribution of spatial units			O		O		●	●			O
A4 Free configuration of the floor plan	O	O	O		●	O	●	●	●	●	O
Separation of support and infill systems											
A5 Open frame structure	O	O	O		●	O	●	●	O	O	●
A6 Independent distribution of services to units		●	O		●		O	●	●	●	O
A7 Access floor or service zones		O			●			●	O	●	●
A8 Infill systems for services	O	O	O		O		O	O	O	O	O
A9 Infill systems for partitions	●				●	●					O
A10 Infill systems for facades	●		O								O
Open Building process											
A11 Distribution between support and infill	O			O							O
A12 Procedures for user participation	●	O					●				
A13 Functional and technical design distinguished	O		O							O	
A14 Implementation of infill unit by unit				●							

● criteria met   O criteria partly met

Fig. 3.4 Characteristics of Open Building. A schema developed by Ulpu Tiuri in Finland. Matrix courtesy of Ulpu Tiuri.

1 Recognizing levels
Applying professional tools and methodologies specifically developed to organize residential work by different parties according to levels and to reorder technical interfaces. Such methods include those developed at the SAR and OBOM in the Netherlands, and at the BRI/MOC and HUDc in Japan.

2 Distributing decision-making
Distributing control for each environmental level to decision-makers on that level.

- Establishing legal, contractual and physical frameworks in which the individual household may design or alter their dwelling unit layout, and determine equipment within their own dwelling.
- Clearly distinguishing collective and individual realms of decision-making, and separating decisions about common spaces and infrastructure from decisions concerning individual dwellings.
- Separating procurement and construction for base building and infill. The latter may then be designed and installed just prior to occupancy on a unit-by-unit basis.
- Avoiding the appropriation of decision-making across several levels by any single entity: in large projects, a single party cannot successfully design the urban tissue, the facades and buildings, the dwellings, and the furniture. Advocates of long-term environmental diversity and health argue that even in smaller multi-family projects, no single entity should exercise multi-level design control.

3 Physically separating environmental levels
- Containing urban infrastructure and making it accessible entirely within public property (where feasible).
- Separating Support from infill
  - constructing Support and infill in two clearly defined steps;
  - placing all components belonging specifically to individual dwellings at the infill level, under direct control of occupants; and

- locating building structure and common mechanical systems infrastructure (building-level cabling, ducts, main supply and drainage piping, and so on) so as to maximize freedom in designing the infill level, while rationalizing connectivity of mechanical systems between base building and fit-out.

4 Coordinating and disentangling subsystems

- Coordinating subsystems for eventual change, thereby allowing them to be independently adjusted or replaced without disrupting other dwellings or subsystems. Using positioning and dimensioning rules such as those based on the 10/20cm band grid developed by the SAR, or the multiple positioning grids developed for the Next21 project in Osaka.
- Selecting 'open' systems with standardized technical interfaces, dimensions and locations, so that any subsystem which adheres to industry-wide performance standards may be used. Choice is then broadly based on design, quality, service and other economic factors, rather than solely on functional compatibility.

5 Enabling household choice and decision-making

- Recasting the role of the dwelling designer as a professional who assists inhabitants in realizing their own dwelling preferences.
- Utilizing information management tools that immediately show dwellers the implications of their design decisions. For example, utilizing software that illustrates the effect of each appliance, system or finish selection on the final installation price of an infill package.
- Supporting and enabling the free configuration of space by tenants.
- Within rental housing, allowing tenants to own and maintain infill within rented space.

6 Using specific Open Building design methods

- Using the *SAR 73* Tissue Method as a means of calculating the infrastructure costs and trade-offs of various tissue models and density criteria.

- Designing base buildings according to books such as *Variations: The Systematic Design of Supports* (N.J. Habraken *et al.*, 1976).

7   *Using specific Support technologies to support the use of infill systems*

- Among the many systems discussed below are: tunnel-formed *in situ* Supports; depressed floor slab trenches; flat beam skeleton; inverted slab-beam floor structures; Z-beam structures; pipe-stairwell organizations; etc.

8   *Using residential infill technologies*

- Installing partial or complete residential infill systems such as Matura, Interlevel, ERA, the KSI Infill System, etc.
- Using RTA (ready-to-assemble) interior systems, particularly partitions, doors, cabinets and other systems or products with high potential for reuse, such as Panekyo products in Japan, IKEA products, Bruynzeel kitchens, etc.
- Employing quick-install door frames and doors that can be installed in under 10 minutes per door.
- Specifying partition systems that are debris-free and easily and quickly installed on-site.
- Superimposing a raised floor above the structural floor on a dwelling-by-dwelling basis.
- Using wiring raceways and quick-connect cabling that allow easy and safe installation and reconfiguration of power and data lines by the user.
- Using pre-terminated cabling such as that manufactured by Wieland, Woertz or National Panasonic.
- Using heating, air conditioning and ventilation equipment designed for efficient distribution of heat and cooling and for efficient installation, energy conservation, and low maintenance; for example, the Sanyo split system, the radiant floor system of Tokyo Gas or the ventilation systems developed for the Esprit Infill System.
- Using advanced plumbing systems such as Delta-Plast or Hepworth piping that includes push-fit (non-solvent welded) fittings for drain

lines and pressurized effluent discharge or macerating water closets. The latter allow discharge into small pipes, eliminating the need for sloped gravity flow solid waste drainage.

9 Using specific OB financial instruments, including those developed in the Buyrent infill purchase system initiated in the Netherlands or the Tsukuba Method form of ownership originated in Japan.

These and similar processes and products allow Open Building to work in practice. They minimize the number of physical interfaces. They also reduce conflict among the various parties involved in designing, installing and maintaining them. Such autonomy of subsystems increases efficiency in construction. It encourages innovation and industrial production while supporting the ongoing globalization of the building industry and its standards.

3.5 OPEN BUILDING STRATEGIES

3.5.1 Basic overview of strategies

Buildings interweave technical products and occupant needs and actions in complex ways. As technical requirements and individual preferences grow more diverse, new ways of working are required to make such increased variety at least as easy to manage as uniformity. The basic physical systems approach in Open Building practice is accordingly to identify, develop or use principles of ordering and combining subsystems (of any scale) by which interference between them – and between the parties controlling them – is minimized.

3.5.1(a) Balancing

From an organizational perspective, Open Building provides tools for professionals to use in distributing responsibility to strike a good balance between overall community coherence and individual freedom in each project. OB helps to identify and assign a clearly defined place and level for action by each stakeholder, and a balance between shorter term individual initiative and variation and long-term community values.

3.5.1(b) Enabling efficiency and variety

Open Building simultaneously enables efficient work processes and variety of physical and organizational patterns in building. These attributes were long presumed to be antithetical in housing construction, although how to combine them is well understood in manufacturing, evident in the 'mass customization' of products as diverse as automobiles and dolls. While optimizing efficiency and systematic production and assembly of facade, roof and structural systems for base buildings, OB also allows each dwelling to be highly customized and changeable.

3.5.1(c) Ordering

Open Building uses specific professional tools and methods to reorganize design work, technical interfaces of installations and regulatory permitting. It employs industrial components in a directed way, according to the principle of levels, for example in the deployment of infill systems. Efficiency results from the combined use of **ordering principles** (rules of three dimensional positioning). Ordering principles minimize interference among subsystems, clearly define interfaces between them and enable unambiguous separation of responsibilities. As a result, Open Building eliminates disruptive reverberation of one part's change through the whole form. The principle of eliminating reverberation can be observed at work in historical built environments in many ways, and on many levels:

- In urban neighborhoods, individual buildings on lots can change or be replaced without forcing an adjustment in overall neighborhood spatial and formal order. This insures the stability and continuity of the urban fabric.
- In buildings, parts that embody long-term physical, cultural and social requirements can remain relatively constant while other parts, including

interior spaces and equipment associated with individual occupancies, change more frequently. Thus, in the Parisian entresol pattern, mezzanines can easily extend or be eliminated in response to changing demands related to production or commerce, without affecting courtyards, entries, concierge stations, neighboring apartments or residential units above. Facade systems that are distinct from the masonry structure (for example, curtain walls) similarly enable component repair and replacement without violating the structural integrity of the base building.

3.5.1(d) Interchangeability

OB's particular approach to interchangeability is based on levels. Within subsystems such as heating and cooling equipment, discrete assemblies can often be replaced by similar products from a variety of manufacturers, just as standardized memory chips from a variety of manufacturers can be installed in most computers. The entire system need not be replaced.

3.5.2 Specific technical strategies

3.5.2(a) Separating base building, infill systems and subsystems

Open Building's primary technical strategy runs directly counter to the goal of total systems integration and unified design control that characterized so much 20th century building research, technology, policy making and ideology. In disentangling subsystems and integrating by level, Open Building argues that cast-in-place-plumbing, integrated prefabricated wall panels and even wiring bundled and buried into walls encumbers the orderly design, installation and upgrading of systems. Implementation of technology otherwise turns dwelling decisions into base building decisions as a byproduct of integration across levels. In the process, control is appropriated upward, onto a higher decision-making level.

In multi-family housing, OB professionals distinguish between spheres of action and responsibility, and the physical elements under the control of each in both the design and placement of systems: there is the collective at various levels (city, neighborhood, apartment complex owner, condominium association or cooperative) and the individual household.

Both Support and infill consist of numerous technical subsystems. In the Support, for example, the facade may be an independent assembly of elements and materials, a highly organized 'kit of parts,' (including conventional prefabricated sun rooms and curtain wall systems). Even when changes in common needs and requirements become apparent, the Support, by virtue of its technically distinct subsystems, can be upgraded with far less disturbance to households than is the case in conventional integrated or 'unibody' construction. Within an infill system, partitioning, cabinetry, appliances and other equipment, when well organized for rapid detachment and replacement, may similarly function as autonomous subsystems with minimal interface entanglements.

3.5.2(b) Disentangling subsystems

Open Building minimizes interfaces and interdependencies among subsystems. Each system is given a dedicated zone and rules of deployment, unlike most conventional construction. Reordering subsystems deployment avoids the confused and chaotic interweaving of piping, wiring and ductwork between or through structural elements in floors and walls. It further avoids the disruptive collision of trades so characteristic of conventional residential construction. As a direct result of disentangling subsystems, initial installation and future renovation and replacement work is streamlined. When every component and run of pipe is installed in a pre-assigned location, quantity surveys become more automated and more accurate. Guesswork is eliminated from current and future renovation processes. In some cases, coordinated infill systems may be certified for installation by a single specially-trained installer, replacing several trades on site.

3.5.2(c) Manufacture and design for free assembly and disassembly

Benefits from exploiting industrial manufacturing capabilities have been realized in many consumer-oriented sectors. In housing, that potential still remains untapped. Product compatibility standards among residential building subsystems made by different manufacturers are nonetheless developing out of necessity. Open Building provides a logical approach to implementing standards. This will eliminate product incompatibility arising out of competing dimensional and performance standards in products with tight interfaces.

The development of 'click-together' and similar components will greatly benefit the housing industry. Such 'open' products with high degrees of compatibility produce higher re-use value. In addition to the obvious efficiencies, cost savings and sustainability advantages of recycling component assemblies with high added value, users consequently will enjoy far greater freedom to safely recombine parts with little or no professional intervention or consequent disruption.

3.5.3 Development strategies

3.5.3(a) Increasing property value while decreasing risk

Rapid deterioration of common building elements and lack of on-site cost control represent fundamental inefficiencies that hamper residential development.

Developers exert considerable effort toward managing both short- and long-term risk. Their strategy is frequently to limit and control risk, while protecting and enhancing the future value of present investments. Despite such efforts, residential buildings become obsolete or require expensive maintenance and renovation. In many nations and markets, the problem is compounded by incentive systems and traditions that encourage short-term investment decisions in buildings, with little value assigned to abating the long-term effects of premature deterioration. Common parts of the building sometimes fail or require rebuilding within a decade; they commonly do so within twenty years of initial construction.

Distinguishing infill from base building can be efficient, and a good investment strategy for large projects, particularly when skilled labor is in short supply and market demand for quality construction is high. Even when enabling individuals' ability to customize units is not a specific goal, separating construction by level provides closer control of the building process, lower on-site labor costs, and generally improved quality. Conversely, when large, complex residential buildings are centrally organized on a single level, project control becomes more complex, driving up the costs of coordination, making quality harder to achieve, and preventing decision deferment that is otherwise highly valued.

3.5.3(b) Deferring investment decisions

In an attempt to accurately predict costs, residential developers and their marketing consultants conventionally require unit floor plans to be designed and most alternatives to be fixed prior to creating the project *pro forma*. This typically occurs several years before final unit rent-up or completion of unit sales. Decision deferment in conventional developments is difficult. Developers therefore greatly benefit from evaluating base building capacity for various alternatives unit floor plan at an early stage, while nonetheless deferring final build-out decisions as long as possible.

3.5.3(c) Improving the climate for developing multi-family housing

In Japan, initiatives such as the Tsukuba Method respond to the prohibitive cost of home ownership and owner disincentives for developing land. There, a 'two step housing process' is coupled with a major rethinking of the land availability problem. In this case, OB objectives include establishing a new form of land ownership and encouraging families to remain in their multi-family dwelling. Dwelling units designed to change according to household life style and cycle represent an integral implementation step toward achieiving those overall goals.

3.5.4 Organizational strategies

3.5.4(a) Disentangling and distributing control

It is normally the case that no single entity can perform all the work efficiently or cost-effectively in large development or construction projects. Independent subcontractors routinely join in the teamwork, and competitive subcontract bidding is generally advantageous. The normal distribution of work responsibility nonetheless causes considerable difficulty; therefore, nimble construction management is very highly valued. In OB projects, the rational distribution of control and responsibility result in reduced conflict and confusion.

3.5.5 Market strategies

3.5.5(a) Delivering consumer preferences through industrialization

Automobiles, stereo equipment, computers, home furnishings and equipment all figure prominently in international consumer culture. Such consumer goods capitalize on industrialized production methods to efficiently provide broad distribution, high quality, systematic variety and competitive choice.

Many trends point to comprehensive dwelling infill systems emerging as the next major consumer product for the housing market. Yet the relationship between industrialization and housing in this century has remained murky, making the path to achieve this circuitous. In almost every country, private and public sector interests periodically call for improvements. These include industrialized production of housing 'units' to boost quantity and quality. Nonetheless, housing has long resisted the sorts of sophisticated production techniques which have become conventional in other industries.

Housing is technically, socially and organizationally complex. It is far more complex than office buildings, which have seen the most advanced infill systems developments; it rivals the complexity of laboratory or hospital buildings. The complexity of housing is rooted in many conditions, including: dwelling preferences; local site characteristics; building regulations and traditions; the social organization of labor; the sheer technical load of utility systems reaching into each space; and the vagaries of real estate investment.

While housing remains invested with local ways of building, it is increasingly influenced by professional and international trends in design, in style, in consumer products and technology. Housing is a bulky consumer-oriented commodity; yet at the same time it embodies a jointly held social asset. That fundamental duality was recognized a third of a century ago in Habraken's first formulation of Supports. That housing has never achieved the straightforward match with consumer preferences achieved by other industrial products is partially a result of that duality.

The alternative to providing consumer choice is to provide limited options: a minimal series of generic fixed floor plans may be designed and engineered. The resulting 'models' are expected to accommodate a broad range of changing needs and preferences of diverse users over time. Deriving the models is time consuming; and it is highly unlikely that one size, or half a dozen, really fits all by the time the project goes to market. Certainly, as market demand, demographics and life styles shift over time, particularly as the population ages, such predetermined models become increasingly out-moded. In conventional approaches to housing multiple families, the under-lying assumption is that households must ultimately adjust to the constraints of settings designed by and for others. At best, the final dwelling in conven-tional practice represents a mutually unsatisfactory compromise between designer, building owner and occupant.

3.5.6 Environmental and sustainability strategies

3.5.6(a) Increasing building life
Because Open Building projects are designed with change in mind, they en-able the retrofitting of outmoded housing stock in an efficient but inhabitant-oriented way, obviating the need for new construction of whole buildings as a result of untimely obsolescence of the building's parts or units.

3.5.6(b) Building for change
Over time, a building typically accommodates diverse uses. Also during cer-

tain parts of its life cycle, multi-family stock experiences frequent occupant turnover. Adaptation in response to changing preferences and technical requirements is easiest when each dwelling is largely autonomous, and each subsystem can be replaced while the main system remains useful.

3.5.6(c) Accommodating natural variety

Steady increase of variety among dwellings over time is a characteristic of historically sustainable built environment. In contrast, conventional multi-family housing now minimizes the number of unit layouts or 'models' to the extent tolerated by the immediate housing market, then sets them in concrete, or binds them in technical entanglement. Their infill is designed and engineered to remain fixed in large measure because the industry has not yet learned to make socially-complex multi-level buildings changeable. Designing, optimizing, financing, engineering, and constructing buildings by following a preestablished mix, layout and placement of individual dwelling types has been easier to accomplish, though it renders even minor changes in dwelling mix or layout problematic.

Uniformity and rigidity do not result from industrialization of subsystems: they rather reflect the centralized organizational structure of large scale housing producers. Open Building allows for design by and for each actual inhabitant rather than for an abstract 'market' (or to meet uniform space and equipment standards). This enables an architecture of coherence coupled with variety in dwelling form, income and household make-up.

3.5.6(d) The long-term and the short-term

The long-term qualities and physical elements of buildings are those which, in most cases, represent long-term community values and investments. They are the parts of the whole which contribute a sustained sense of place and environmental coherence. Short-term physical elements represent more individual values. They reflect the preferences, concerns and investments of individuals or individual organizations. They are, by definition shorter term investments, elements which wear out sooner or must be upgraded more frequently. Achieving sustainability in a 'throw-away culture' of immediacy begins with understanding the relative life spans of building parts and sets of building elements which belong together.

3.5.7 Coordination strategies

3.5.7(a) Minimizing reverberation and conflicts in coordination

Open Building's coordination strategies are founded upon basic principles and observations: Buildings and their ability to transform are shaped by technical, constructional and social control hierarchies and dependencies: For example, a building's load-bearing interior walls are dependent on walls or columns below: no walls beneath them must be shifted. In terms of constructional hierarchy, cast-in-place electrical conduits and boxes must be fully in place before the concrete is poured. In terms of social hierarchy, if heating pipes for an entire building pass through private dwellings, the building's management must maintain the right to access those pipes; if electrical lines for multiple occupancies pass behind the gypsum board surface of interior partition walls in a multi-family condominium, it is prudent to state in the by-laws that nothing may be nailed or bolted to the walls.

In all three instances – technical, constructional and social – built-in technical conditions will limit tenant freedom throughout the life of the building. Such limitations are intensified when techniques from freestanding dwelling design and construction – such as running building structure, pipes and conduit throughout floor and wall cavities and interstitial floor spaces – are carried over into multi-family housing. The result is a coordination or interface problem in all three spheres.

Open Building minimizes unintentional control hierarchies that create reverberation within a given level or between levels, while minimizing conflicts in coordination. Minimizing reverberation between adjacent units ensures that new construction and renovation of individual units will affect only the Support or the individual unit under construction. To further reduce reverberation between subsystems within the unit, assembly in correct sequence, modular coordination and the use of standard interfaces are required.

Minimizing coordination conflicts among multiple trades and interest groups within the dwelling or at the level of the Support means that the logistics of on-site delivery, construction sequencing, definition of interlocking scopes of on-site work and sequencing must be rationalized. Materials must arrive on site only at the right time. They must be easily assembled by a minimal number of distinct and well-coordinated trades and contractors, who minimize the number of sequential visits. Reducing the scale of each project – coordinating teams to build one dwelling at a time within the Support, rather than an entire floor – also prevents logistical problems. Specifically, change in the infill of one unit is prevented from reverberating down through all the trades.

3.6 SUMMARY

The quick overview of Open Building presented in Part One has constituted a primer on the subject. An extensive general reading list for more in-depth examination appears at the conclusion of this Part.

The history, theory, research and practice of Open Building is diverse and evolving. Because residential OB brings together agents from many different disciplines who build in many diverse settings, it does not present a unified set of principles, beliefs, goals or technologies. Nonetheless, for different reasons, residential structures throughout the world are beginning to be constructed in similar ways. Above, we have outlined the commonalities in goals and approaches, methods and strategies that exist between the various communities of Open Builders.

In Part Two, we present case studies of projects that represent important milestone developments toward residential Open Building over the past 35 years.

REFERENCES

1. A critical history of the SAR and its role in applied housing research is scheduled for fall 1999 publication: Bosma K. (ed), van Hoogstraten, D. and Vos, M. *Housing for the millions: John Habraken and the SAR 1960–2000*, NAi Publishers, Rotterdam.
2. Included among the SAR's projects was preliminary development of computational tools for the design of Supports, one of the first computer aids to residential design (Gross, 1998)
3. 'Yuumeiku,' the Japanese name coined for the project, derives from the multiple meaning of 'you make' (do-it-yourself) + 'yume' (dream) + 'iku/ikiru' (to realize, grow or live).
4. In the majority of S/I projects realized to date, the public agency building the Support has also supplied the infill.
5. In the United States, public housing interventions have always represented a relatively small percentage of total housing output.
6. For further detailed discussion of levels, see Habraken (1998).
7. *SAR 65* presented a method for systematically evaluating proposed designs in terms of capacity. Later this was given detail for practice in Habraken *et al.* (1976) *Variations: The Systematic Design of Supports*. Habraken' s Tools of the Trade: Thematic Aspects of Designing (1996) further clarifies the distinction between function and capacity.

ADDITIONAL READINGS

Basic texts

Bosma, K. (ed), van Hoogstraten, D. and Vos, M. (1999) *Housing for the Millions: John Habraken and the SAR, 1960-2000*. NAi, Rotterdam.

Carp, J. (1985) *Keyenburg: A Pilot Project*. Stichting Architecten Research, Eindhoven.

Dluhosch, E. (principal researcher) (1976) *IF (Industrialization Forum); Systems Construction Analysis Research*. Published jointly at Montréal, Harvard, MIT and Washington University. 7 no. 1.

Habraken, N.J. (1999) *Supports*. Second English Edition. (ed J. Teicher) Urban International Press, London.

Hamdi, N. (1991) *Housing Without Houses: Participation, Flexibility, Enablement*. Van Nostrand Reinhold, New York.

Kendall, S. (ed). (1987) Changing Patterns in Japanese Housing. Special issue, *Open House International.* **12** no. 2.

Proveniers, A. and Fassbinder, H. (n.d.) *New Wave in Building: A flexible way of design, construction and real estate management.* Eindhoven University of Technology, Maastricht, Netherlands.

Turner, J.F.C. (1972) Supports and Detachable Units. Special Edition, *Toshi-Jutaku.* September.

Turner, J.F.C. (1979) Open Housing. Special Edition, *Toshi-Jutaku.* January.

van der Werf, F. (1980) Molenvliet-Wilgendonk: Experimental Housing Project, Papendrecht, The Netherlands. *The Harvard Architectural Review.* **1** Spring.

Wilkinson, N. (ed). (1976-present) *Open House International.* London.

Further readings

Andrade, J., Santa Maria, R. and Govela, A. (1978) Transformacion de un Entorno Urbano: Santa Ursula 1950–1977. *Architectura y Sociedad.* **1** no. 1.

Bakhtin, M.M. (1981) *The Dialogic Imagination: Four Essays by M.M.Bakhtin.* (ed M. Holquist, and trans C. Emerson and M. Holquist). University of Texas Press, Austin, Texas.

Beisi, J. and Wong, W. (1998) *Adaptable Housing Design.* Southeast University Press, Nanjing.

Brand, S. (1994) *How Buildings Learn: What happens after they're built.* Viking, New York.

Cuperus, Y. and Kapteijns, J. (1993) Open Building Strategies in Post War Housing Estates. *Open House International.* **18** no. 2. pp. 3–14.

Fukao, S. (1987) Century Housing System: Background and Status Report. *Open House International.* **12** no. 2. pp. 30–37.

Gann, D. (1999) *Flexibility and Choice in Housing.* Policy Press, UK.

Gross, M. (1998) Computer-Aided Design. *The Encyclopedia of Housing* (ed Van Vliet) Sage, Thousand Oaks, Calif.

Habraken, N.J. (1964) The Tissue of the Town: Some Suggestions for Further Scrutiny. *Forum.* **XVIII** no. 1. pp. 23–27.

Habraken, N.J. (1964) Quality and Quantity: the Industrialization of Housing. *Forum.* **XVIII** no. 2. pp. 1–22.

Habraken, N.J. (1968) Housing – The Act of Dwelling. *The Architect's Journal.* May. pp. 1187–1192.

Habraken, N.J. (1968/69) Supports: Responsibilities and Possibilities. *The Architectural Association Quarterly.* Winter. pp. 29–31.

Habraken, N.J. (1970) *Three R's for Housing.* Scheltema en Holkema, Amsterdam.

Habraken, N.J. (1972) Involving People in the Housing Process. *RIBA Journal.* November.

Habraken, N.J. (1971) You Can't Design the Ordinary. *Architectural Design.* April.

Habraken, N.J. (with Boekholt, Thyssen and Dinjens). (1976) *Variations: The Systematic Design of Supports.* MIT Press, Cambridge, Mass.

Habraken, N.J. (1980) The Leaves and the Flowers. *VIA, Culture and the Social Vision.* MIT Press, Cambridge, Mass.

Habraken, N.J. (1986) Towards a New Professional Role. *Design Studies,* **7** no. 3. pp. 139–143.

Habraken, N.J. (1988) *Transformations of the Site.* Awater Press, Cambridge, Mass.

Habraken, N.J. (1994) Cultivating the Field: About an Attitude when Making Architecture. *Places.* **9** no. 1. pp. 8–21.

Habraken, N.J. (1996) Tools of the Trade. Unpublished lecture, MIT Department of Architecture.

Habraken, N.J. (1998) *The Structure of the Ordinary: Form and Control in the Built Environment.* (ed J. Teicher). MIT Press, Cambridge, Mass.

Habraken, N.J. with Aldrete-Haas, J.A., Chow, R., Hille, T., Krugmeier, P., Lampkin, M., Mallows, A., Mignucci, A., Takase, Y., Weller, K., and Yokouchi, T. (1981) *The Grunsfeld Variations: A Demonstration Project on the Coordination of a Design Team in Urban Design.* MIT Laboratory for Architecture and Planning, Cambridge, Mass.

Hasegawa, A. (ed). (1994) Next21. Special issue, *SD (Space Design) 25.*

Herbert, G. (1984) *The Dream of the Factory-Made House: Walter Gropius and Konrad Wachsmann.* MIT Press, Cambridge, Mass.

Kendall, S.H. (1986) The Netherlands: Distinguishing Support and Infill. *Architecture.* October. pp. 90–94.

Kendall, S.H. (1988) Management Lessons in Housing Variety. *Journal of Property Management.* September/October. pp. 22–27.

Kendall, S.H. (1990) Shell/Infill: A Technical Study of a New Strategy for 2x4 Housebuilding. *Open House International,* **15** no. 1. pp. 13–19.

Kendall, S.H. (1993) (with MacFadyen, D.) Marketing and Cost Deferral Benefits of Just-in-Time Units. *Units.* March. pp. 37–41.

Kendall, S.H. (1993) Open Stock. *The Construction Specifier.* May. pp. 110–118.

Kendall, S.H. (1993) Open Building: Technology Serving Households. *Progressive Architecture.* November. pp. 95–98.

Kendall, S.H. (1994) The Entangled American House. *Blueprints.* **12** no. 1. pp. 2–7.

Kendall, S.H. (1995) Developments Toward Open Building in Japan. Silver Spring, Md.

Kendall, S.H. (1996) Open Building: A New Multifamily Housing Paradigm. *Urban Land.* November. p. 23.

Kendall, S.H. (1996) Europe's Matura Infill System Quickly Routes Utilities for Custom Remodeling. *Automated Builder.* May. pp. 16–18.

Lahdenperä, P. (1998) *The inevitable change: Why and how to modify the operational modes of the construction industry for the common good.* The Finnish Building Center, Helsinki.

Pehnt, W. (1987) *Lucien Kroll: Buildings and Projects.* Rizzoli, New York.

Tatsumi, K. and Takada, M. (1987) Two Step Housing System. Changing Patterns in Japanese Housing (ed S. Kendall). Special issue, *Open House International.* **12** no. 2. pp. 20–29.

Tiuri, U. (1998) Characteristics of Open Building in Experimental Housing. *Proceedings/Open Building Workshop and Symposium.* (ed S. Kendall). CIB Report Publication 221, Rotterdam.

Trapman, J. (1957) Tall Constructions in Oblong Blocks. *Bouw.* March 15. Also (1964) *Forum.* **4.**

Turner, J.F.C. (1977) *Housing by People: Towards Autonomy in Building Environments. New York*: Pantheon Books.

Utida, Y. (1977) *Open Systems in Building Production.* Shokoku-sha Publishers, Tokyo.

Utida, Y. (1994) Aiming for a Flexible Architecture, *GA Japan 06.* January–February.

Utida, Y., Tatsumi, K., Chikazumi, S., Fukao, S. and Takada, M. (1994) Osaka Gas Experimental Housing Next21. *GA Japan 06.*

Utida, Y., Tatsumi, K., Chikazumi, S., Fukao, S. and Takada, M. (1994) Next21. Special Issue, *SD (Space Design).* no. 25.

Ventre, F.T. (1982) Building in Eclipse, Architecture in Secession. *Progressive Architecture.* **12** no. 82. pp. 58–61.

Vreedenburgh, E. (ed). (1992) *Entangled Building...?* Werkgroep OBOM, Delft.

Yagi, K. (ed.) (1993) Renovation by Open Building System. *Process Architecture 112: Collective Housing in Holland.* September.

PART TWO

A SURVEY OF MILESTONE PROJECTS

4

Case Studies

1966   Neuwil
Wohlen, Switzerland

Fig. 4.1 Photograph courtesy of Roger Kaysel.

ARCHITECT: Metron Architect Group
OWNER: Housing cooperative
DWELLINGS: 49 rental units
SUPPORT CONSTRUCTION: Eight-story concrete slab and column; fixed stairs,
    bathrooms and kitchens
INFILL PROVISION: Demountable interior walls

This eight-story apartment building contains 49 rental units with 'flexible' interior space divisions. Dwelling dimensions are standardized to one fixed size. Sizes, locations, and products for stairs, kitchens and bathrooms are also fixed. All units are oriented east-west.

Dwellings are accessed via a common central corridor. The bathroom and kitchen of each unit are located in the interior, which has no direct natural illumination or ventilation. The interior spaces adjacent to facades are identically sized and have identical balcony spaces. The east-west orientation of units assures that both front and back receive ample sunshine. Because the quality, size and solar orientation of these spaces is essentially identical, the living room can face either direction.

Interior layout of the dwellings may be determined by the tenants and changed according to their preferences. Space is designed to be partitioned according to a 30cm grid, using any of five varieties of ready-made wall panels. All five kinds of gypsum board panels are stored in a common room in the building, available for tenant use. They are made in 60cm or 90cm widths and are light and easily moved. To assist in this process, the architects prepared *My Flat is My Castle*, a three-part users' manual with easy-to-read design drawings and illustrations:

Part One: The history of a family searching for their own castle. This section describes the changing needs and changing floor layouts of a hypothetical family inhabiting this building for more than ten years. The text uses descriptions, sketches, and photographs of apartment models.

Part Two: The Apartment Division Guide. This section introduces inhabitants to the wall elements, how to assemble them, and cost management strategies.

Part Three: Apartment Division Floor Plans. Sample floor plans illustrate a wide range of possibilities for space division. Each example includes a short description of the client household and characteristics of the spatial functions. In each floor plan, heavy lines designate the Support, which tenants may not move or modify. Thin lines indicate furniture and movable walls, and dotted lines show the possible positions for walls.

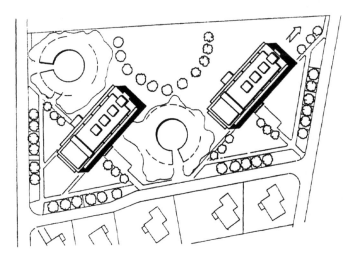

Fig. 4.2 Site plan. Drawing by Hans Rusterholz and Alexander Henz, courtesy of Metron Architects.

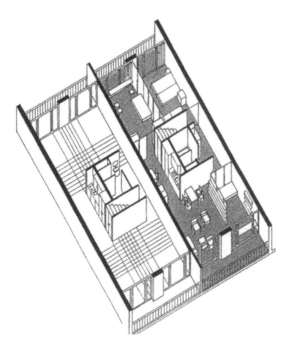

Fig. 4.3 Optional dwelling unit wall positions and example of a dwelling unit. Drawing by Hans Rusterholz and Alexander Henz, courtesy of Metron Architects.

Fig. 4.4 Residents installing an interior partition. Photograph courtesy of Metron Architects.

1974   Maison Médicale student housing ('La Mémé')
Catholic University of Louvain, Brussels, Belgium

Fig. 4.5 Photograph courtesy of the Office of Lucien Kroll.

ARCHITECT: Atelier d'Urbanisme, d'Architecture et d'Informatique Lucien Kroll
OWNER: Catholic University of Louvain
DWELLINGS: 20 apartments, 60 studios, 200 rooms for single students, 200 single
   rooms grouped into apartments, six communal houses of 18 rooms, and
   social spaces.
SUPPORT CONSTRUCTION: Concrete slab and column, demountable curtain wall
   facade, some electric cabling and piping in the concrete slabs
INFILL PROVISION: Demountable walls

The student residences form part of a larger 40 000m² complex of buildings designed by
Lucien Kroll. The total project includes married student housing, religious facilities, a
restaurant, a primary school, a theater and an underground rail station. When the
Catholic University of Louvain decided to move its medical facilities from Louvain to
Brussels, the students and their organization, La Maison Médicale, engaged the office of
Lucien Kroll. Kroll and his team were invited to design the social zone with the direct
participation of clients and inhabitants. The architects sought to maximize differentia-
tion between dwellings, to avoid repetition, and to preserve a sense of *genius loci*.

   The appearance of disorder in the highly organized and modular building plan and
facade is somewhat deceptive. It is entirely coordinated according to the SAR 10/20cm
'tartan' grid. Load bearing elements and fixed equipment are located in the 20cm bands.

Partitions and other detachable elements are positioned in the 10cm bands. The structure employs flat slab construction, with 'wandering columns' placed at multiples of 90cm. Columns are positioned away from the facade, leaving it free of structural elements. According to Kroll, the columns form a 'mosaic of square or rectangular umbrellas which support each other at the edges . . . regular columns support conformity, while irregular ones stimulate the imagination.' Electrical conduits, plumbing and heating pipes are set within the (slightly thickened) slab. The structure was built to endure, while the infill was expected to require cyclical updating or replacement. Accordingly, the infill is removable, and consists of manufactured products.

Partitions, which are movable, are made of gypsum board sheets glued to a core of mineral wool, making them both insulating and also self-supporting without the need for posts. They are easily installed because of the flat concrete ceilings. Jacks hold the panels against the ceiling, allowing them to be erected and relocated without the help of professionals. Windows, including their frames, are sized according to the basic module of 30cm. The frames are of different colors in order to accentuate the specific identities of the various components. Sanitary equipment and kitchens are grouped and fixed as part of the Support.

Fig. 4.6 Photograph courtesy of the Office of Lucien Kroll.

Fig. 4.7 Detail of the facade.
Photograph courtesy of the Office of
Lucien Kroll.

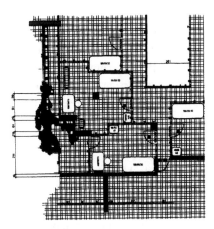

Fig. 4.8 Detail of the building plan
showing 10/20cm SAR grid. Drawing
courtesy of the Office of Lucien Kroll.

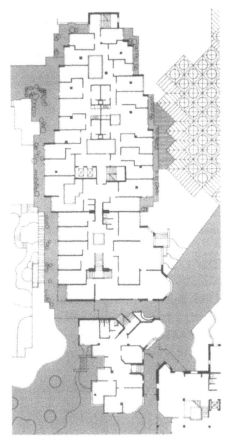

Fig. 4.9 Site plan. Drawing courtesy of
the Office of Lucien Kroll.

1976   Dwelling of Tomorrow
Hollabrunn, Austria

Fig. 4.10 Photo courtesy of the Office of Architect Professor Ottokar Uhl.

ARCHITECT: Dirisamer, Kuzmich, Uhl, Voss and Weber
OWNER: Non-profit Housing Association
DWELLINGS: 70 units
SUPPORT CONSTRUCTION: Concrete panel system (required in the competition)
INFILL PROVISION: Normal interior construction

This 'Dwelling of Tomorrow' competition entry was awarded first prize by the Austrian Ministry of Housing and Technology in 1971. The project was completed in 1976. Open Building methods were used to assist in the planning, design, construction and project delivery processes. SAR methodologies also aided communication as the traditional roles of the parties involved were redefined. The participants in the project included politicians, financiers, and professionals, as well as the users themselves, who participated in all phases.

A number of non-standard conditions were established for the project. These included:

1. Enabling delayed decisions – occupants need time to decide; future alterations must be possible; dwelling size must be able to change over time.
2. A new type of sale or rental contract would establish the size and location of the units, but not their plan layouts.

3. The ability to calculate individual tenant costs was established as a prerequisite for participation of the prospective dwellers.
4. Information and consultation guidance was required concerning construction and equipment choices as well as alternative space planning layouts.
5. Occupants had the right to take part in planning and to direct the designers.
6. Joint Administration: tenant participation would not end with project completion; occupants would maintain the right to assume management control of the cooperative over time.

User participation started with the beginning of construction. Regular meetings were held with users, architects and representatives of the Housing Association. During the meetings, future households received detailed information concerning dwelling unit type, size, layout possibilities, costs, construction schedules and so on. To aid in the ongoing deliberations at home, each household was provided with a blank floor plan showing only the Support and the positions of vertical service elements. Examples of possible floor plans were provided only on request. As a result, floor plans and facades are different for each dwelling unit. During construction, an on-site scale model display was kept current, allowing each household to see their unit in the context of the whole project.

Fig. 4.11 Study model of the project. Photo courtesy of the Office of Architect Professor Ottokar Uhl.

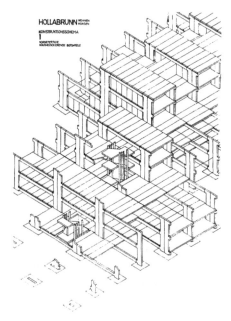

Fig. 4.12 Schematic diagram of the
Support structure. Drawing courtesy of
the Office of Architect Professor
Ottokar Uhl.

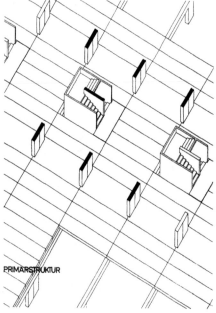

Fig. 4.13 Schematic drawing of the
primary structure. Drawing courtesy of
the Office of Architect Professor
Ottokar Uhl.

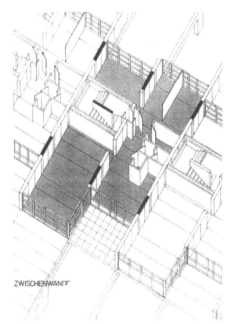

ZWISCHENWAND

Fig. 4.14 Schematic drawing of
the Support parcellation (subdivision).
Drawing courtesy of the Office of
Architect Professor Ottokar Uhl.

EINRICHTUNG

Fig. 4.15 Schematic drawing of the
infill of several units. Drawing courtesy
of the Office of Architect Professor
Ottokar Uhl.

1977 Beverwaard Urban District
Rotterdam, Netherlands

ARCHITECT: RPHS Architects
DWELLINGS: 5000 units (about 12 000 people)

This development of 157 hectares of farm land south of Rotterdam was laid out according to the principles of *SAR 73*. The plan was to develop an urban district including approximately 5000 units of housing and associated public facilities, stores and offices. To coordinate decision-making among the many players, several 'tissue models' were devised and adopted.

The town plan was designed to heighten social contact, and the relation of places of activity to defined spaces. The intent was to heighten the experience of a number of spatial qualities: identity, intimacy, protection. The project's large scale is minimized, bringing most dwellings into direct contact with the ground. All dwellings are variations within several well-established vernacular themes, adhering to detailed performance specifications regarding the public-private relationship. Several spatial themes, rather than any programmed functions, thus provided the starting point for urban design.

First these spatial themes were defined at the level of the land use plan, specifying the relative positions of buildings and open spaces. Only then were 'thematic' as well as 'non-thematic' functions positioned. Various building functions (shopping areas, offices, schools, etc.) and open space functions (parking, main and secondary streets, parks, etc.) were then distributed within the established spatial structure.

To coordinate the work of all architects involved in the project and to achieve a balance between overall coherence and local variation, distinct tissue models with coordinated plan dimensions were developed. For the first time, three-dimensional specifications regulating urban built form were legally described in drawn documents, rather than in words only. A number of independent architects subsequently designed portions of the whole tissue. Each section has its own style, program, dwelling sizes, characteristics, and details, in accordance with the rules of the urban tissue.

In its completed form, Beverwaard is a continuous urban tissue of boulevards, streets, squares, alleys, canals, gateways, courtyards, parks and so on. Dwellings, shops, offices and other normal town functions are not prescriptively separated, they are interwoven within a continuous built fabric.

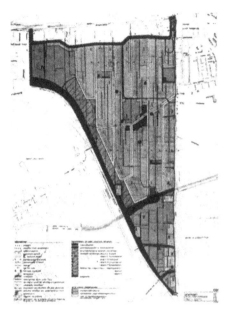

Fig. 4.16 Site plan of district south of Rotterdam. Drawing courtesy of RPHS Architects.

Fig. 4.17 Illustrative site plan detail: tissue models adjusted to the site. Drawing courtesy of RPHS Architects.

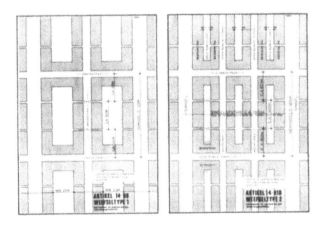

Fig. 4.18 Two tissue models showing basic block types and dimensional rules. Drawing courtesy of RPHS Architects.

1977　Sterrenburg III
Dordrecht, Netherlands

Fig. 4.19　Photo courtesy of De Jong and Van Olphen, Architects.

ARCHITECT: De Jong and Van Olphen
OWNER: Dordrecht-Zwijndrecht Housing Association
DWELLINGS: 402 units
SUPPORT CONSTRUCTION: Tunnel-formed cast-in-place concrete; prefabricated wood
　　frame facade units
INFILL PROVISION: Bruynzeel Infill System

This project was designed to maximize user participation at the request of the multi-party client, which included a housing association and two municipalities. A Support was designed to accommodate 402 units of housing organized in two categories: medium -high-rise units (121) and terrace or row house units (281). The terrace house units were further divided into three types: dwellings with pitched roofs, with transverse roofs, and with asymmetrical roofs. All were of the same plan dimension: 9.6m deep by 5.4m wide.

Interconnecting the blocks resulted in a variety of profiles, permitting varied physical planning solutions. In addition, the roof form variation allowed some variation in unit size. The medium-high-rise dwellings also comprise various types, organized in a staggered terrace form. In all unit types, the Support includes the position of internal stairs, utility meters, vertical service shafts, and fixed dimension openings in the facade walls.

Ten alternative plan layouts were developed by the architects. In what subsequently became standard practice, one was designated as a 'standard version,' on the basis of which a base dwelling price was determined. Prices for variant layouts were obtained by adding or subtracting the appropriate amount. The ministry that offered subsidies also calculated the annual rent rebate and the total costs of the standard unit.

The Support has finished walls, ceilings, floors, gables and roofs, with standardized openings for utility shafts and stairs. The floors include an extra thick concrete topping, which contains wiring conduits and central heating pipes attached to radiators in individual dwelling units. Many design and technical decisions and details enable easy extension of the units. To make the separation of Support and its infill economically feasible, the high investment in infill and coordination costs was balanced by savings in labor during Support construction.

The assembly kit for the infill comprises the partitions, doors, bathroom and kitchen and most of the mechanical equipment installations. It consists of an entirely prefabricated kit-of-parts, including wooden partition framing, framing connector blocks, and surface panels. Surface-mounted electrical raceways were also used where wiring had to run along walls rather than in the floor conduits. Modular coordination was used to assure compatible products and interfaces. To avoid the normal 'parade' of specialized trades roaming through the site multiple times, a single multiple-trade team was provided by Bruynzeel to install the infill.

Fig. 4.20 Site model of the project.
Photo courtesy of De Jong and Van Olphen,
Architects.

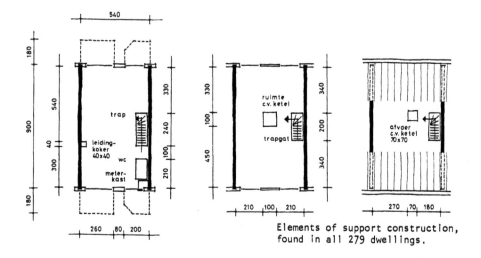

Elements of support construction, found in all 279 dwellings.

Fig. 4.21 Plans of the Support. Drawings courtesy of De Jong and Van Olphen, Architects.

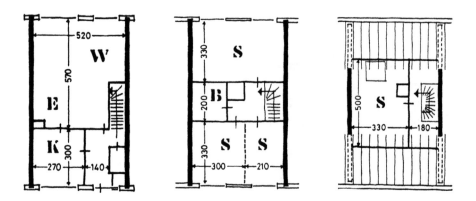

Fig. 4.22 Dwelling unit variants. Drawings courtesy of De Jong and Van Olphen, Architects.

1977 Papendrecht
Molenvliet, Netherlands

Fig. 4.23 Photograph courtesy of John Carp.

ARCHITECT: Frans van der Werf, Werkgroep KOKON
OWNER: Housing Association of Papendrecht
DWELLINGS: 124 rental dwellings; 4 office spaces
SUPPORT CONSTRUCTION: Tunnel-formed cast-in-place concrete, with
    openings in slabs for shared vertical mechanical systems and internal stairs;
    kit-of- parts facade.
INFILL PROVISION: Conventional Dutch interior construction.

The winner of a competition for 2800 dwellings at a density of 30 dwelling units/
hectare, this project won on the combined merits of its urban design, architecture and
participatory decision-making process. It is organized on four environmental levels:
overall urban plan; tissue (urban design) plan; Support, and infill. Some of its basic
design concepts were based on Christopher Alexander's *A Pattern Language*.

The project's 124 realized dwellings surround courtyards in two-to-four-story
blocks featuring steeply pitched roofs. Most units are entered via one courtyard, with

back yards or roof terraces which open onto another adjacent court. All courts are closed to vehicular traffic.

The Support consists of a highly uniform cast-in-place concrete framework, with openings in the slabs for vertical mechanical chases and stairs within individual units. To allow for variation and changeability in unit designs, the location of Support elements was determined by a series of capacity studies. Tunnel forms – reusable steel forms put in place and moved by cranes – were used in constructing the Support. The concrete walls between bays of the building are regularly spaced, making construction fast and efficient, while still allowing a wide variety of unit configurations. A prefabricated wooden facade framework – an updated version of the typical medieval Dutch canal house facade comprised of a series of joined wooden frames – was installed as part of the Support.

Infill for each unit was determined after arranging units of required floor area, or 'parcelling out' (parcellation) of the Support. This process, like the subsequent fitting out, involved user participation. Each household met individually on several occasions with the architect, progressing from rough sketches to final drawings. Once they had been signed by the occupants, final drawings were translated into construction documents. The infill of each unit included interior walls, doors, trim and finishes; bathroom cabinets and equipment; kitchen cabinets and fixtures; electrical and mechanical equipment for the unit; closets; and windows and doors inserted into the Support facade framework.

The project incorporates many traditional elements of Dutch urban housing – pitched roofs, wooden windows, doors opening onto courtyards and some mixed use: interspersed among the dwellings are doctors' offices, small shops and commercial offices, even a motorcycle parts shop. This project demonstrated that even in multi-family housing, the variation characterizing households can be expressed easily and beneficially on the exterior of the building. In this case, inhabitants worked with designers to determine window frame color and arrangement to complement the custom interior layout of each unit.

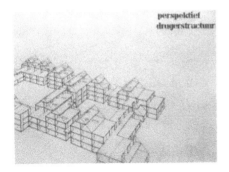

Fig. 4.24 The Support structure. Drawing courtesy of Frans van der Werf.

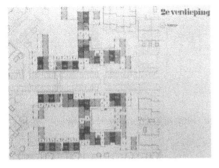

Fig. 4.25 Second floor of the Support showing dwelling parcellation. Drawing courtesy of Frans van der Werf.

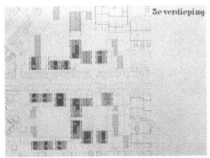

Fig. 4.26 Third floor of the Support showing dwelling parcellation. Drawing courtesy of Frans van der Werf.

Fig. 4.27 Aerial view of the project. Photograph courtesy of Michel Hofmeester, AeroCamera BV.

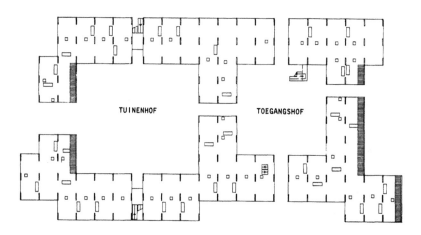

Fig. 4.28 Support before dwelling parcellation. Drawing courtesy of Frans van der Werf.

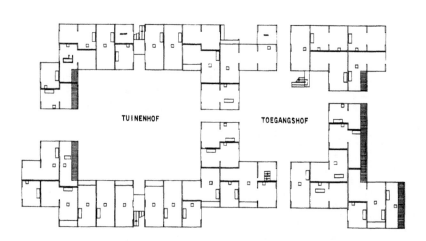

Fig. 4.29 Support after dwelling parcellation. Drawing courtesy of Frans van der Werf.

AUTOSTRAAT

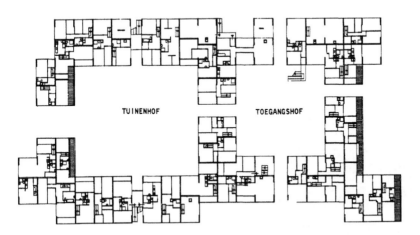

TUINENHOF   TOEGANGSHOF

Fig. 4.30 Support showing dwelling infill. Drawing courtesy of Frans van der Werf.

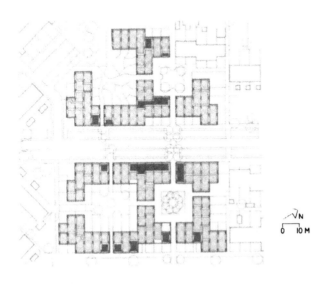

Fig. 4.31 Site plan showing courtyards, parking and non-residential uses. Drawing courtesy of Frans van der Werf.

1979   PSSHAK/Adelaide Road
London, England

Fig. 4.32 Photograph by Stephen Kendall.

ARCHITECT: Greater London Council (GLC) Architecture Department; Nabeel Hamdi
   and Nicholas Wilkinson, architects-in-charge.
OWNER: Greater London Council (original owner)
DWELLINGS: 45 units
SUPPORT CONSTRUCTION: Concrete slabs bearing on perimeter walls of brick/block
   cavity construction; central district heating
INFILL PROVISION: Bruynzeel component system

PSSHAK stands for Primary Systems Support and Housing Assembly Kits. In proposing an adaptable and flexible approach, the PSSHAK projects offered an alternative to
standard methods of housing that would more closely match the needs of tenants than
was possible under normal GLC regulations.

'Stamford Hill,' as the first PSSHAK project came to be called (after its London
neighborhood), was completed in London in 1976. Adelaide Road, the second PSSHAK
project, followed three years later. Located in Camden, London, it comprises a 45-

unit Open Building cluster. Despite many planning and construction innovations, the detailing and general appearance of both schemes were somewhat straightforward in appearance, albeit non-traditional.

The eight three-story buildings of the Adelaide Road project are served by off-street parking. Apartments are directly entered at the ground level or accessed via public stairs serving galleries at upper stories. The Support is conventionally built of concrete slabs and piers with brick veneer. Portions of the slab are left open to accommodate vertical mechanical equipment services, to house internal stairs for bi-level units, or to be filled-in by wooden floor panels. The Support includes all exterior doors and windows, public stairs, the roof and the main mechanical systems. Each Support was designed for a range of unit sizes, which can be combined for a total of between 64 standard units (one- and two-bedroom units) and 32 larger units.

During the project's programming and design phases, the housing authority select-ed 45 tenant households according to their standard procedure. Tenants in groups of 12 met with the architects to receive orientation regarding the process and the assembly kit. They were then given two weeks in which to create a first design of their own units. These sketches were reviewed and finalized with the architect acting as a 'skilled enabler.' The manufacturer's representative worked with the architect to keep each unit's kit of parts within the budget allowance for each tenant. The assembly elements were then prepared, delivered and assembled in a completely 'dry' process.

For dwelling infill, the project used the PSSHAK kits developed by the architect team, notably including a kit of parts assembly provided by Bruynzeel BV in the Nether-lands. The Bruynzeel infill included a prefabricated interior partition system, bath units, kitchen elements, electrical and mechanical systems, doors, trim and finishes.

Savings in design and construction time resulted from simplification of pre-contract and construction procedures. Overall project cost was only slightly higher than for conventional planning and construction. Subsequent cost savings in modernization are presumed to be significant. Subsequent tenants rarely exploited the units' capacity for easy transformation. Nonetheless, surveys of successive generations of tenants con-sistently revealed very high levels of satisfaction. The project was recently privatized, at which point tenants were offered subsidies to purchase their dwellings as condominiums.

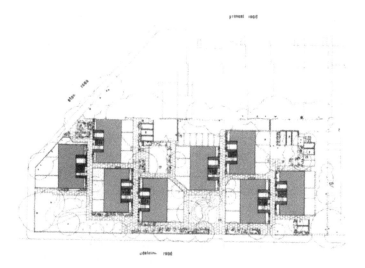

N

Fig. 4.33 Site plan. Note off-street parking and walkways between buildings. Drawing courtesy of Nabeel Hamdi.

Fig. 4.34 Model of the Support and alternative infill layouts. Photo courtesy of Nabeel Hamdi.

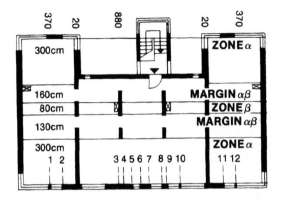

Key:

Black line denotes Support structure.

Hatched line denotes demountable party wall.

Open line denotes 'kit' (infill components).

Numbers (1,2,3...12) indicate possible positions of partitions.

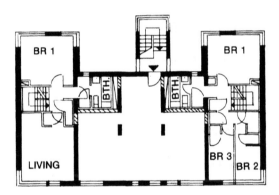

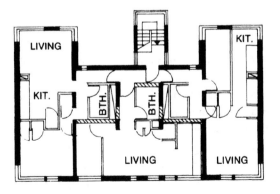

Fig. 4.35 Typical floor of the Support and alternative dwelling unit arrangements. Drawings courtesy of Nabeel Hamdi.

1979 Hasselderveld

Geleen, Netherlands

Fig. 4.36 Photograph courtesy of Bert Wauben Architects.

ARCHITECT: Bert Wauben
OWNER: Geleen Non-profit Housing Association
DWELLINGS: 71 units
SUPPORT CONSTRUCTION: Concrete frame, brick veneer
INFILL PROVISION: Conventional interior construction

This project aimed in its planning to realize dwellings of above-average quality with both internal and external adjustability. The architect's studies of Pompeii found expression in a site plan incorporating a pattern of passages and patios. The patio of each dwelling provides a focus for living and bedrooms; it is also the 'margin' for possible future expansion of each dwelling.

The overall site layout places a green zone incorporating children's play facilities at the project's center. Many dwellings are situated immediately around this green space. The remaining dwellings are located within the restricted-traffic precinct in various courts, each of which is connected to the central green space by pedestrian passageways. Even with these amenities, the project achieves a higher density than traditional plans.

The dwellings were planned to enable variations in stacking during the design phase. The one-story unit was designed to be stacked in blocks of from two to four staggered stories, providing 12, 14 or 15 dwellings. In this project, one block of two stories, one block of three stories, and three blocks of four stories were built, combining for a total of 71 dwelling units. Such 'patio bungalows' provide an alternative to traditional terraced housing. Access to the dwellings on the second level is via a ramp: in all, 60 of the 71 units are accessible on grade. Living rooms and bedrooms are clustered around a patio to give a high level of intimacy. Different 'minimum dwelling' variants are available, with one, two or three bedrooms; linear, L-shaped or Z-shaped living rooms; and so on. Sixty-four different layouts from which to choose were developed.

Over the years, inhabitants have transformed many individual dwelling layouts and facades as well as exterior spaces. Some units have been extended into the courtyards, as anticipated.

Fig. 4.37 Photograph courtesy of Bert Wauben Architects.

Fig. 4.38 Photograph courtesy of Bert Wauben Architects.

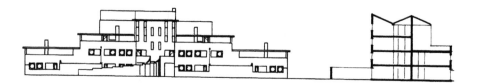

Fig. 4.39 Building elevation and cross section. Drawing courtesy of Bert Wauben Architects.

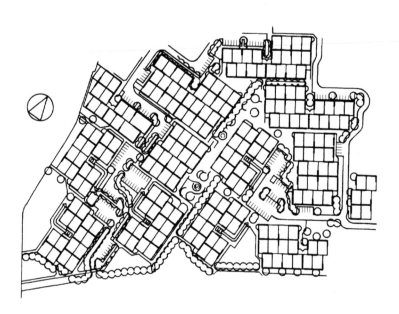

Fig. 4.40 Site plan. Drawing courtesy of Bert Wauben Architects.

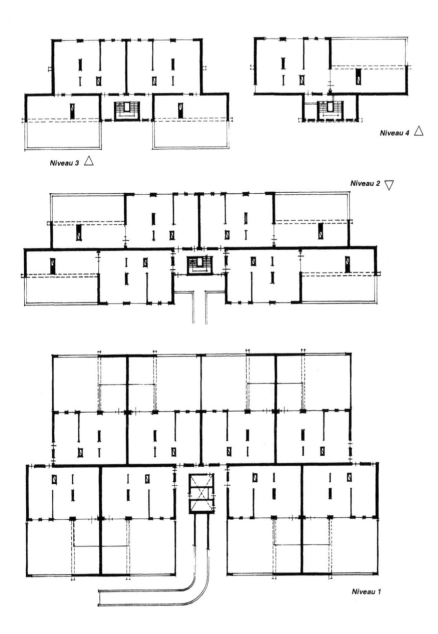

Niveau 4

Niveau 3

Niveau 2

Niveau 1

Fig. 4.41 Floor plan of the Support from ground level to upper floors. Drawing courtesy of Bert Wauben Architects.

1983   Estate Tsurumaki and Town Estate Tsurumaki
Tama New Town, Japan

Fig. 4.42 Estate Tsurumaki. Photograph courtesy of Seiichi Fukao.

ARCHITECT: HUDc and Kan Sogo Design Office + Soken Assoc. + Alsed
    Architectural Laboratory
OWNER: Housing and Urban Development corporation
DWELLINGS: Estate Tsurumaki, 190; Town Estate Tsurumaki, 29
SUPPORT CONSTRUCTION: Concrete cast-in-place slab and cross walls
INFILL PROVISION: Unit bathrooms; raised floor; movable walls; movable storage cup
    boards; conventional electric distribution

These two condominium projects were among the first practical application projects
developed as a result of the KEP (Kodan Experimental Project), an initiative begun in
1974. (2.2.3)

Estate Tsurumaki is a series of four-story walk-up buildings, containing dwelling
units ranging in size from 87–89m². The Housing and Urban Development Corporation (HUDc) offered a large variety of preliminary fixed unit plans in each building.
After move-in, occupants were able to change layouts using movable partitions and storage units. A 1997 research survey revealed that changes in household composition
account for many changes observed in a number of dwellings. In other cases, families
changed layouts inherited from the previous owners.

In the immediately adjacent Town Estate Tsurumaki project, HUDc subsequently
developed 29 bi-level townhouse units ranging in size from 99–105m², with units

grouped in blocks of two to four units. In all the units, buyers could select from six different ground floor layouts that were fully predetermined by HUDc. For the upper floor, HUDc offered three kinds of dwelling choices, the 'All-Free,' 'Semi-Free,' and 'All-Set' types.

In the 'All-Free' type, the upper floor was left entirely open. Except for an installed toilet room, there were only unpainted walls. Buyers were free to arrange and subdivide the space. In the 'Semi-Free' type, half of the upper floor was completed by HUDc, and the remaining half left for owner design. In the 'All-Set' type, the entire upper floor was determined by HUDc, and finished by the supplier.

Further choices and upgrades were also available. For example, kitchens could be standard or could be upgraded. Center for Better Living (BL) certified closet units could be selected and installed. Various finishes could be selected from a large menu of options. Finally, buyers could opt for a solar collector to be placed on the roof.

Fig. 4.43 Site plan showing both project phases. Drawing courtesy of HUDc.

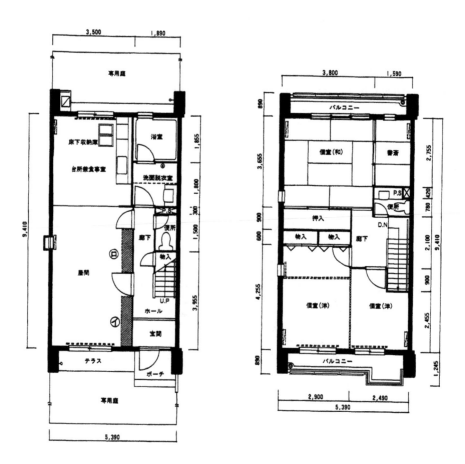

Fig. 4.44 Unit plan variations. Drawing courtesy of HUDc.

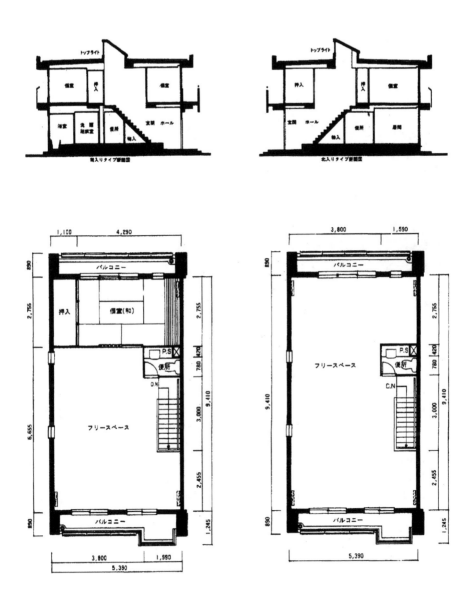

Fig. 4.45 Building cross sections and unit plan variations. Drawing courtesy of HUDc.

1984   Keyenburg
Rotterdam, Netherlands

Fig. 4.46 Photograph by Stephen Kendall.

ARCHITECT: Frans van der Werf, Werkgroep KOKON
OWNER: Tuinstad Zuidwijk Housing Association
DWELLINGS: 152
SUPPORT CONSTRUCTION: Tunnel-formed cast-in-place concrete; openings for piping
INFILL PROVISION: Nijhuis 4DEE system, surface-mounted electric raceways.

A large housing association, Tuinstad Zuidwijk, was interested in exploring a new way to build and manage housing units in Keyenburg, a district of Rotterdam. In this project, rent levels are adjusted for the first time according to the total amenity package specified by tenants. Incentives for tenants to reduce costs and increase personal responsibility were actively created. The project was designed to attract a mix of ages and incomes; and also to retain and accommodate current neighborhood residents who wanted smaller apartments.

The project consists of four buildings, four stories each, surrounding a large central green space. Buildings facing the main street have ground floor commercial lease space, while ground floor units on side streets have street-level apartments. The Support allows for variation in unit size. 115 two-person units, 32 one-person dwellings and 5 units for handicapped people were fitted out. Access to upper level dwellings is provided by an exterior gallery on each floor, served by an elevator and stairs, and wide enough for seat-

ing and plantings initiated by households. Window frame colors were selected by each
household from a color pallet provided by the architect.

The Support construction is tunnel-formed concrete, with a brick veneer facade
over an insulated, prefabricated wooden facade frame. Vertical piping and mechanical
system shafts occur in each bay, their position optimized according to an analysis of the
capacity for varied dwelling unit layouts. Infill uses the 4DEE system of Nijhuis, including a bathroom placed on a raised floor to allow adequate horizontal drain piping runs
to the vertical plumbing stack, thus permitting the bath unit to be somewhat freely
located. Keyenburg was one of a number of selected trial projects for the proposed
national modular coordination standard based on SAR studies. That standard was subsequently adopted in the Netherlands.

The design process used by client and architect was similar to that used in other
Open Building projects by the architect. The housing authority provided Van der Werf
with a list of interested and eligible tenants, each of whom was asked to specify their preferred location in the Support. Aided by a full scale mock-up, future tenants laid out
their own unit plans in sketches, specifying some finishes and other details. The architect digitized the sketches and rendered them within a computer software program.
Output based on optimized material take-offs immediately informed tenants precisely
how their choices would raise or lower the monthly rent, based on a standard fit-out
price. Approved design revisions including changes in amenity choices could be assessed
on the spot. The same computer program then produced more detailed technical drawings and material quantity surveys based on the final approved design.

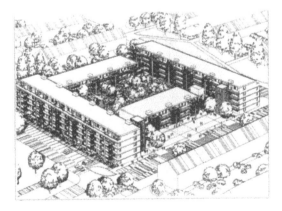

Fig. 4.47 Axonometric aerial view of the project. Drawing courtesy of
Frans van der Werf/Werkgroep KOKON.

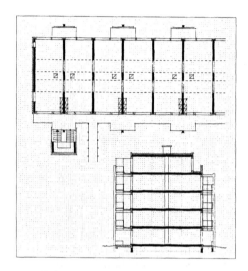

Fig. 4.48 Support plan and section. Drawing courtesy of Frans van der Werf/Werkgroep KOKON.

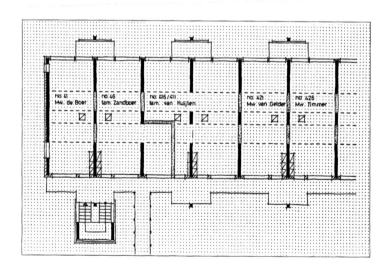

Fig. 4.49 Support parcellation (subdivision). Drawing courtesy of Frans van der Werf/Werkgroep KOKON.

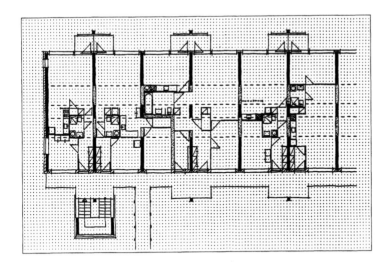

Fig. 4.50 Drawing showing the infill for each unit. Drawing courtesy of Frans van der Werf/Werkgroep KOKON.

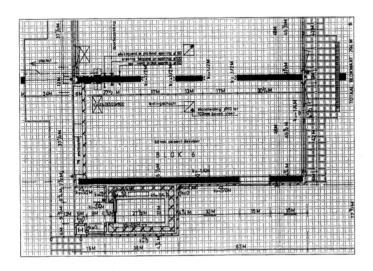

Fig. 4.51 Technical drawing of one bay of the Support with 10/20cm tartan grid. Drawing courtesy of Frans van der Werf/Werkgroep KOKON.

1985   Free Plan Rental
Hikarigaoka, Tokyo, Japan

Fig. 4.52 Photograph courtesy of Seiichi Fukao.

ARCHITECT: HUDc and Kan Sogo Design Office
OWNER: Housing and Urban Development corporation
DWELLINGS: 30 rental units
SUPPORT CONSTRUCTION: Rigid concrete frame; piping trenches in slab
INFILL PROVISION: Traditional interior construction

While the Century Housing System (CHS) was under development, HUDc began 'Free Plan Rental,' an experiment inspired by Support/Infill housing implemented in the Netherlands. Two projects were realized, one in Tokyo in 1985, and another in 1988 in Tama New Town.

The first project, in Hikarigaoka in Tokyo, consists of 30 dwelling units ranging in size from 61.5–71.5m². Of 500 families applying to participate at project commencement, 30 were selected. HUDc owns the site, the Support and the common piping. Tenants rent space but own the infill, including all partitions, finishes and mechanical equipment. The kitchen and sanitary spaces must be located near fixed pipe shafts. However, utilizing the piping trench allows the toilet to be located up to 1.5m from the pipe shaft. Therefore, a wide variety of plans has occurred.

Of the three dwelling options, the 'Free Space Type' enables dwellers to select the entire infill, for which they are also responsible. In the second, or 'Semi-Free Space

Plan,' dwellers may select only that portion of the infill to be freely determined. The 'Menu Select Type' allows dwellers to select from a limited menu of options.

HUDc also prepared three options for the second Free Plan Rental project, located in Tama New Town. In that project, if the renter buys 'standard plan' infill from among the models offered, HUDc constructs it at a fixed price. Or, HUDc introduces the renter/buyer to an infill contractor, who then individually contracts to construct custom infill, following design manual guidelines established by HUDc. The third option is a completely Do-It-Yourself (DIY) approach, in which HUDc involvement is limited to presenting the design manual to be followed. When an occupant moves out, HUDc buys the infill or helps sell it to the new dweller according to a depreciation schedule and certain rules regarding the rental contract, the design and the infill components used.

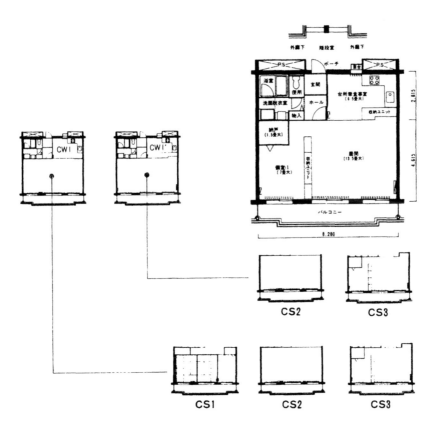

Fig. 4.53 Menu selection of dwelling units. Drawing courtesy of HUDc.

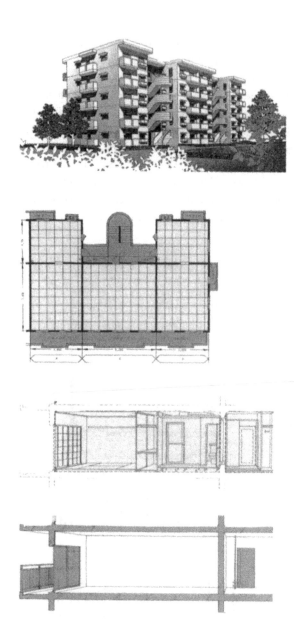

Fig. 4.54 Perspective view, plan and section diagrams of the Support and infill. Drawings courtesy of HUDc.

1987 Support Housing, Wuxi
Hui Feng Xin-Cun, Wuxi, China

Fig. 4.55 Photograph courtesy of Bao Jia-sheng.

ARCHITECT: Bao Jia-sheng and Wuxi Housing Bureau
OWNER: Wuxi Housing Bureau
DWELLINGS: 214
SUPPORT CONSTRUCTION: Concrete hollow-floor planks on masonry bearing walls;
   kitchens and baths fixed as part of the Support
INFILL PROVISION: Dwellers made their own infill using available conventional
   products.

This experimental Support housing project in Wuxi was the first of its kind in China. Its
primary aims included developing user participation in the housing process and study-
ing new ways to make housing that can adjust over time to accommodate changing
household needs. The project resulted from collaboration between the Center for Open
Building Research and Development (COBRD) in Nanjing, and the Wuxi Housing
Management Bureau. COBRD was responsible for the site and architectural design; the
Wuxi Housing Bureau developed it and also provided engineering for structural, electrical and mechanical systems.

The project is composed of eleven buildings: nine 'set-back' courtyard types and
two 'villa' types. The courtyard type has four model plans, and the three-story villa
buildings offer a choice between two plans. Eighty-three percent of all dwellings are
within four floors of the ground. The average dwelling unit floor area is 55.76m². The

building construction utilizes traditional brick bearing walls, hollow-core concrete planks for the floor structure and traditional tile roofs. Interior fit-out is traditional.

Each building is organized by a basic 'Unit Support,' a Z-shaped spatial unit which basically corresponds to the domain of a single household. Within the Unit Support, the designers freely located all collective, private and utility spaces. The Unit Supports were thus composed in a large variety of arrangements. Public stairs and 'plug-in-units' were also added, each in itself capable of a large number of variants in internal spatial arrangement. The 11 buildings are massed in stepped form around courtyards. They present an image of orderly variety with traditional Chinese architectural motifs.

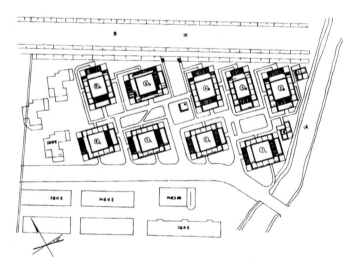

Fig. 4.56 Site plan. Drawing courtesy of Bao Jia-sheng.

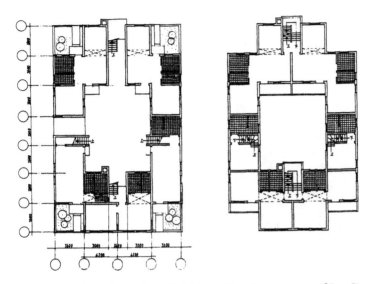

Fig. 4.57 Floor plans of a typical block. Drawing courtesy of Bao Jia-sheng.

Fig. 4.58 Building section of a terraced block. Drawing courtesy of Bao Jia-sheng.

1989　Senri Inokodani Housing Estate Two Step Housing
Osaka, Japan

Fig. 4.59 Photograph courtesy of Osaka Prefecture Housing Agency.

ARCHITECT: Osaka Prefecture Housing Agency + Tatsumi/Takada + Ichiura Architects
OWNER: Osaka Prefecture Housing Agency
DWELLINGS: 33 units
SUPPORT CONSTRUCTION: Concrete slab with concrete shear walls; depressed slab in
　　the middle of each bay
INFILL PROVISION: Raised floor, prefabricated partitions, unit bathrooms

This public housing project combines the Two Step Housing Supply and Century Housing Systems. The Two Step approach enables the public sector to play a guiding role, while recognizing the importance of private initiative. In this project, the Support was built as 'social overhead capital,' common property characterized by good quality and long durability. Fit-out of the Support was the second step. In this particular project, the public agency supplied both the Support and the infill. Nonetheless, the two were kept physically distinct to ensure future ease of modification.

　　The Century Housing System combines modular coordination, a planning grid for partition location, and the concept of assembling component groups according to the anticipated 'durable years' of each. To accommodate the relatively limited durability of mechanical equipment and piping, new coordination principles were developed to guide interfaces between component groups and the Support.

The two buildings in the project are five and six stories, respectively. They contain 33 dwelling units averaging 103m². Units are paired on each floor around combined stair and elevator cores. Luxury units with roof terraces occupy the top floors. The Support uses shear walls with openings between the structural bays, rather than columns. A utility trench is located in the middle zone of each unit, where kitchen, bath unit and toilet wet cells can be positioned with some variation. The other two zones contain living spaces whose layouts are also variable. A raised floor is used throughout the units.

The use of a utility trench is common to many CHS projects. To date, design, contracting and fit-out of these projects have not been performed by independent suppliers. Nonetheless, the potential for independent fit-out is inherently built in; and in lending themselves to long-term adaptability, these Supports achieve a major goal of Open Building.

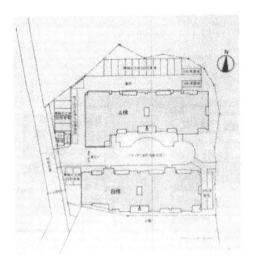

Fig. 4.60 Site plan. Drawing courtesy of Ichiura Architects.

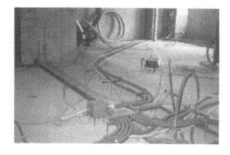

Fig. 4.61 Service lines being installed in Support trench. Photograph courtesy of Mitsuo Takada.

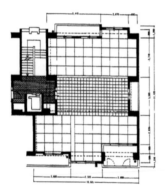

Fig. 4.62 Support plan. Drawing courtesy of Ichiura Architects.

Fig. 4.63 Support and alternative dwelling units. Drawing courtesy of Ichiura Architects.

1990– Patrimoniums Woningen/Voorburg Renovation Project
Voorburg, Netherlands

Fig. 4.64 Building before and after Support renovation. Note new units on ground level. Photographs courtesy of Karel Dekker.

ORIGINAL ARCHITECT: Lucas & Neimeyer
RENOVATION ARCHITECT: RPHS Architects
OWNER: Patrimoniums Woningen Housing Corporation
DWELLINGS: 110
SUPPORT CONSTRUCTION: Concrete slab; masonry bearing walls; wood frame facades
    with windows
NEW INFILL: Matura Inbouw, ERA Infill

Patrimoniums Woningen, a large private housing association, owns a property containing many five-story multi-family buildings near Rotterdam. In 1988, the association decided to rationalize management of the property and begin to upgrade it. They decided to modernize the housing stock by renovating vacant residential rental units on a one-unit-at-a-time, as-needed basis, a significant departure from the normal approach of vacating an entire building and upgrading it all at once. At the same time, economic and facilities management analyses pointed to the need to begin a long-term upgrade of the entire site. Included were base building improvements – adding elevators and balconies, replacing the original stairs and mechanical systems – and adding new storage sheds. The owner also decided to add two-story townhouses at the corners of larger apartment blocks, closing the inner courtyard spaces and creating a sense of security and privacy. The original sidewalk-accessible storage rooms on the ground level were replaced with entry level apartments for the elderly and the handicapped.

Matura Infill Systems, a specialized interior fit-out company, was initially contracted to provide dwelling unit infill. During the two weeks required to gut each vacated unit, the new tenant met with the architect. A floor plan and equipment and finish specifications were selected from among several options. The architect's drawings were then transmitted to Matura. One month after being vacated, the unit was again ready for occupancy, with an entirely new interior reflecting the new tenant's preferences. (Chapter 7)

Subsequently, many other inhabitants have decided to modernize their rented apartments. Units are modernized one at a time, each with a custom design. Several different infill systems are currently available to tenants. Depending on the infill provider, dwellings can be installed in ten working days or less. Tenants are simply assessed modest additional monthly fees if the equipment and finishes selected exceed the standard adopted by the building owner. When an occupant moves out, the housing corporation helps to sell the infill to the new dweller, or buys and stores or reinstalls it.

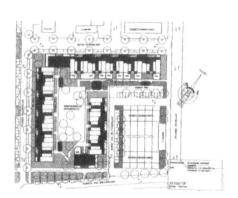

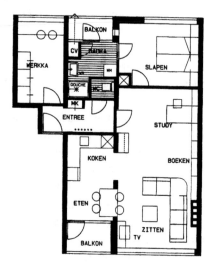

Fig. 4.65 Site plan showing new two-story units at the ends of the large blocks. Drawing courtesy of RPHS Architects.

Fig. 4.66 One dwelling unit plan variant, using Matura Infill System. Drawing courtesy of RPHS Architects.

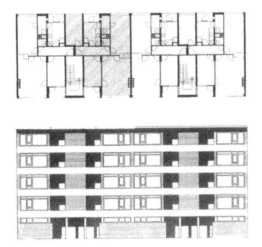

Fig. 4.67 Typical building plan and elevation prior to Support preparation. Drawing courtesy of RPHS Architects.

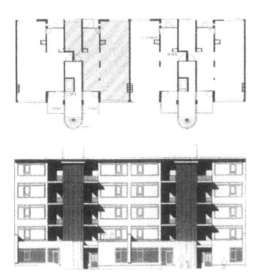

Fig. 4.68 Typical building plan and elevation after Support renovation. Drawing courtesy of RPHS Architects.

1991 'Davidsboden' Apartments
Basel, Switzerland

Fig. 4.69 Interior courtyard. Photograph courtesy of Michael Koechlin.

ARCHITECT: Erny, Gramelsbacher and Schneider, Architects
OWNER: Christoph-Merian-Stiftung
DWELLINGS: 154
SUPPORT: Reinforced concrete elevator building; vertical mechanical shafts
INFILL PROVISION: Interior partitions, bathrooms and kitchens are infill elements

This building, owned and managed by Christoph-Merian-Stiftung (CMS), offers adaptability in two respects: First, units on the same floor can be combined. Secondly, internal partition walls, kitchens and bathrooms are variable and can be changed in relation to fixed vertical mechanical shafts.

The decision to make an adaptable building preceded initial building programming, as did decisions to create a way of living by combining individual needs with collective living; to accommodate present and future quality standards; and to establish a tenant self-management system. Prior to construction, two social workers established an office to help tenants create a tenant association and a self-management system and to serve as an information center.

The first tenants were encouraged to begin to design the layouts of their flats on their own. Specific zones of the dwelling where tenants could make decisions were noted in the leases. Six months prior to the projected date for moving in, the construction manager intervened to enable a formal design process. Each group of ten households held five consultative meetings, using models, floor plans and materials samples to discuss the quality of the building, the varieties of floor plans and the method of tenant participation.

Tenants with a common building entry comprise an association. Tenant associations jointly manage the building, setting rules, maintaining public space and the common heating system. The association also performs small repairs, manages and oversees larger building projects and identifies prospective tenants when a vacancy occurs.

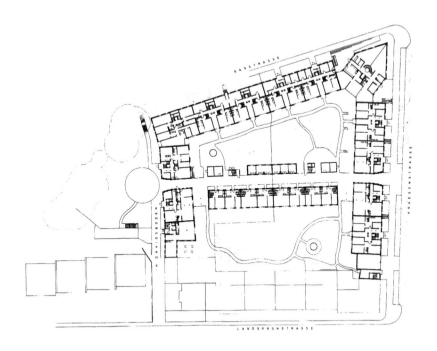

Fig. 4.70 Ground level plan. Drawing courtesy of Erny, Gramelsbacher and Schneider, Architects.

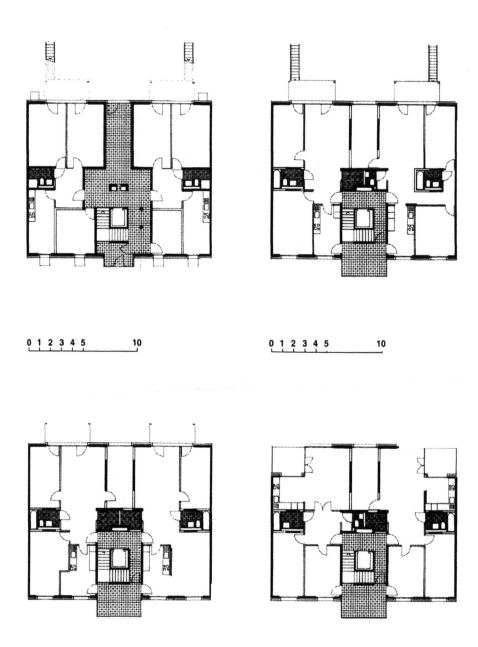

Fig. 4.71 A variety of dwelling units. Drawing courtesy of Erny, Gramelsbacher and Schneider, Architects.

1993   Green Village Utsugidai
Hachioji, Japan

Fig. 4.72 Photograph courtesy of HUDc.

ARCHITECT: HUDc and Han Architects (base building)
OWNER: Green Village Utsugidai Condominium Association
DWELLINGS: 76 condominium units
SUPPORT CONSTRUCTION: Reinforced concrete; piping trench within the slab
INFILL PROVISION: Haseko Corporation

This coop project was built to accommodate varied unit sizes and layouts. It has 76 units ranging in size from 97–173m². Design was organized among three teams of professionals, two of which worked with three-generation households, and the other with general household types. The three-generation-household dwellings were among the first contemporary ones of their kind in Japan.

The Support design was first completed, with occupants jointly determining the design of the common room and exterior layout. Each resident then worked with an architect to make interior layout, equipment and finish decisions. Occupants could design their dwelling's exterior facade in part, following certain rules. Dwellings could also have either one or two entry doors. After all design decisions were completed, the contractor (Haseko Corporation) constructed the building as a whole.

This project employed the principle of a broad, unit-wide 'wet trench' (20cm deep x 300cm wide) for the first time. This piping trench is located between party walls; kitchen, bath unit and toilet pipes must be positioned directly above it. A 75cm x

270cm vertical piping and ventilation shaft is positioned in this trench zone either on a party wall or in the center of a dwelling. Finish floor level is approximately 6cm above the concrete slab (26cm above the trench floor), using a raised floor system supported on pedestals.

Construction of the Support and the infill were not organized in separate contracts. Although a single contractor executed the project, the design process and construction will enable the building to behave as an S/I project in the future, extending its long-term adaptability and predicted useful life.

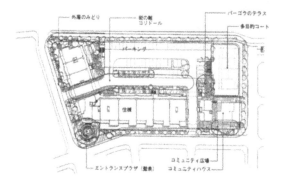

Fig. 4.73 Site plan courtesy of HUDc.

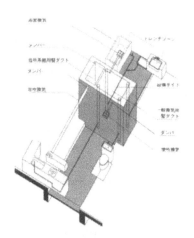

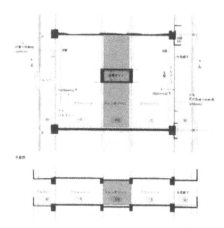

Fig. 4.74 Support service trench. Drawing courtesy of HUDc.

Fig. 4.75 Building plan and section. Drawing courtesy of HUDc.

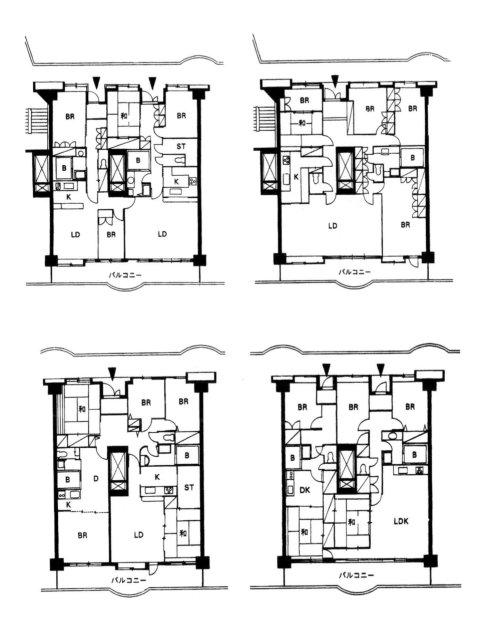

Fig. 4.76 Four different dwelling units.
Drawings courtesy of HUDc.

1994 Banner Building
Seattle, Washington, USA

Fig. 4.77 Photograph courtesy of James Frederick Housel.

ARCHITECT: Weinstein Copeland Architects
OWNER: Banner Building Condominium Association
DWELLINGS: Two penthouses, 11 two-story residential, two one-story residential,
    3 retail/commercial, 5 custom craft, 1 apartment sold as a condo (3 low income
    rentals, 3 market rate rentals)
SUPPORT CONSTRUCTION: Reinforced concrete slabs and columns; common hydronic
water system for heating and cooling in party walls
INFILL PROVISION: Conventional construction

The Banner Building was conceived by Artist/Industrial Designer Koryn Rolstad as a
catalyst for neighborhood revitalization. It was constructed in 1994 on a steep slope in a
deteriorated area near the Seattle waterfront. It was developed to allow residents to purchase undeveloped space as condominium owners – allowing for individual design
development of live/work spaces. Units were then built-out individually, with the main
building performing as a common, serviced structure. The architect created an owner's

manual featuring minimum 'build-out' requirements to be followed during construction. Base building construction cost was modest, $5.8 million, or $65.00/ft^2 ($700/m^2).

The project contains three main kinds of units (residential condominiums, commercial/retail condominiums, and residential rental units). There are 14 two-story live/work dwellings (1800ft^2 or 167m^2) plus an additional 6 rental units (600 and 1200ft^2, or 56 and 112m^2) in a separate freestanding two-story wood frame structure. The remainder of the units are zoned for retail, light manufacturing and custom craft commercial use. The main building is a cast-in-place concrete slab-and-frame structure. It places dwellings on a 'plaza garden' level, situated above a two-story base which houses commercial space and parking garage. The plaza or 'courtyard' between the two buildings is landscaped and maintained by unit owners. Units are accessed via exterior corridors that are eight feet (2.44m) wide. This allows inhabitants to install plantings. It also provided generous emergency egress, as required by the fire code. The second floor of each condominium unit was designated as a mezzanine level, reached by a custom-designed and freely located stair inside each dwelling. The mezzanine can be extended or reduced in size without affecting code requirements.

Plumbing stacks are positioned within parallel, opposing party walls. Bathrooms and kitchens can thus be positioned in a variety of places along those walls. The party walls are of double layered construction with insulation to reduce noise travel. All other 'fit-out' requirements were the responsibility of the owners. Owners have produced a wide variety of layouts. Some units have been reconfigured upon resale.

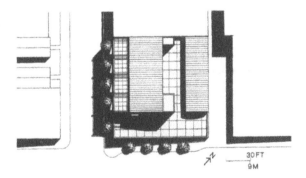

Fig. 4.78 Site plan. Drawing courtesy of Weinstein Copeland Architects.

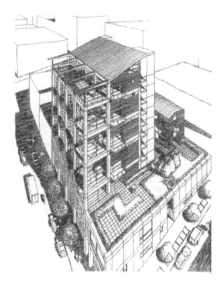

Fig. 4.79 Bird's eye view of the project. Drawing courtesy of Weinstein Copeland Architects.

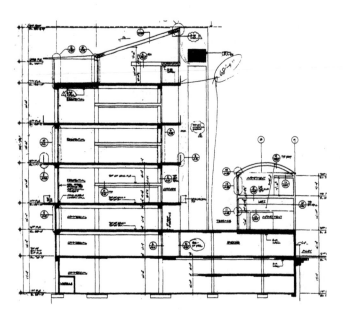

Fig. 4.80 Building cross section. Drawing courtesy of Weinstein Copeland Architects.

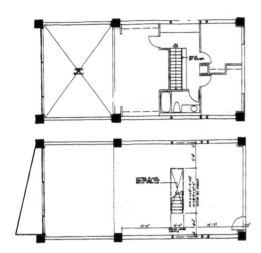

Fig. 4.81 Typical two-story dwelling unit floor plans. Drawing courtesy of Weinstein Copeland Architects.

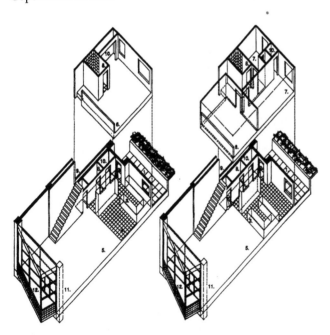

Fig. 4.82 'Exploded' view of typical dwelling units. Drawing courtesy of Weinstein Copeland Architects.

1994 Next21
Osaka, Japan

Fig. 4.83 Photograph courtesy of Next21 Committee.

PLANNING/DESIGN: Osaka Gas and Next21 Planning Team (Utida, Tatsumi, Fukao, Takada, Chikazumi, Takama, Endo, Sendo)

BUILDING ARCHITECT: Yositika Utida, Shu-Koh-Sha Architecture and Urban Design Studio

CONSTRUCTION: Ohbayashi Corporation

DESIGN SYSTEM PLANNING: Kazuo Tatsumi, Mitsuo Takada

DWELLING DESIGN RULES: Mitsuo Takada, Osaka Gas, KBI Architects and Design Office

MODULAR COORD. SYSTEM: Seiichi Fukao

OWNER: Osaka Gas Corporation

DWELLINGS: 18

SUPPORT CONSTRUCTION: Reinforced concrete skeleton; newly developed facade system

INFILL PROVISION: Experimental systems

Next21 is an experimental 18-unit housing project. It anticipates the more comfortable life urban households will characteristically enjoy in the 21st Century. The project was conceived by Osaka Gas Company in collaboration with the Next21 planning team. The Next21 Construction Committee developed the basic plan and design. Its objectives included:

- using resources more effectively through systematized construction;
- creating a variety of residential units to accommodate varying households;
- introducing substantial natural greenery throughout a high-rise structure;
- creating a wildlife habitat within urban multi-family housing;
- treating everyday waste and drainage on site within the building;
- minimizing the building's compound burden on the environment;
- using energy efficiently by means including fuel cells; and
- making a more comfortable life possible without increasing energy consumption.

Units were designed by 13 different architects. Each unit's interior and exterior layout was freely designed within a system of coordinating rules for positioning various elements. The generous floor-to-floor height allowed for the introduction of utility distribution space above ceilings and under raised floors; therefore, ducts and piping can be routed independently of structural elements. Main beams have reduced depth midspan. This allows ducts and piping to pass over the beams without use of 'sleeves.' The main horizontal utility zones are located under exterior corridors or 'streets in the air.'

The building frame ('skeleton'), exterior cladding, interior finishes, and mechanical systems were designed following CHS principles: as independent building subsystems, each anticipates a different repair, upgrade and replacement cycle. Design of the 18 units began after design of the skeleton and continued during its construction. Dwellings and their mechanical systems were designed prior to design of the base building's mechanical system. Subsequently, mechanical services at all levels were installed by a single contractor.

Next21 was constructed as a whole, but designed in such a way that its various subsystems can be adjusted with improved autonomy. To test this objective, one 4th-story unit has been substantially renovated. All work was accomplished from within the unit, using hanging scaffolding, thus minimizing disruption to abutting inhabitants. A substantial percentage of the materials removed – especially of the facade – were successfully redeployed. The project continues to explore new methods for building urban housing and experimental infill systems, to accommodate varying lifestyles with reduced energy consumption. The second phase of Next21 includes renovating other units, introducing a new group of inhabitants, and continued evaluation of the energy systems.

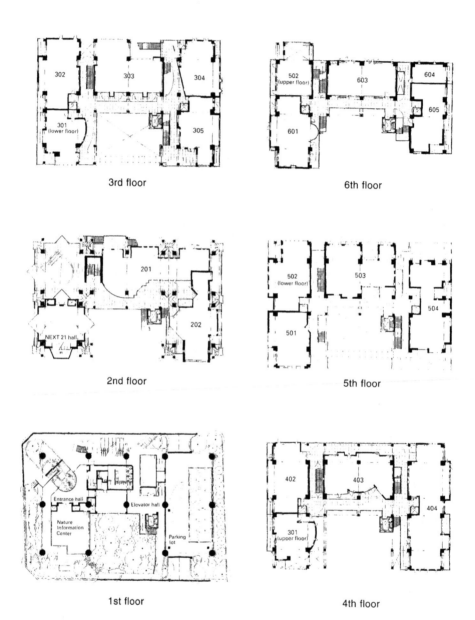

Fig. 4.84 Floor plans. Drawings courtesy of Shu-Koh-Sha Architecture and Urban Design Studio.

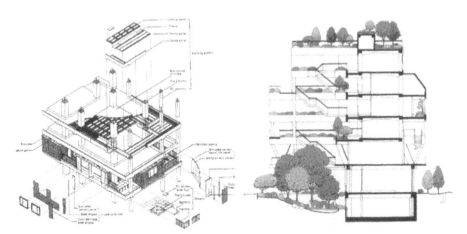

Fig. 4.85 Building systems. Drawing courtesy of Shu-Koh-Sha Architecture and Urban Design Studio.

Fig. 4.86 Building section showing urban natural habitats. Drawing courtesy of Shu-Koh-Sha Architecture and Urban Design Studio.

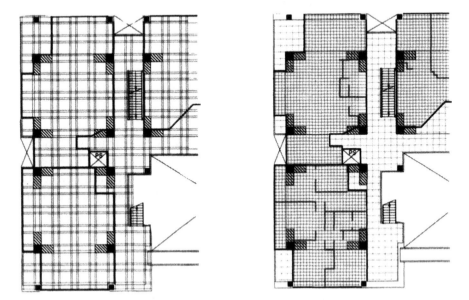

Fig. 4.87 Modular grids for coordination. Drawings courtesy of Seiichi Fukao.

1994　Pipe-Stairwell Adaptable Housing
Cuiwei Residential Quarter, Beijing, China

Fig. 4.88　Photograph courtesy of Zhang Qinnan.

ARCHITECT: Ma Yunyue and Zhang Qinnan, M & A Architects and Consultants
　　International Co., Beijing
OWNER: Leal Housing Technology Development Center
DWELLINGS: 9
SUPPORT CONSTRUCTION: Concrete plank floors and concrete columns with additional
　　masonry bearing walls; vertical mechanical equipment in stairwell space.
INFILL PROVISION: A variety of experimental infill systems

This project, a three-story walk-up building with a stucco exterior finish, was commissioned by the Ministry of Construction as part of the National Building Research Program of China's Eighth Five-Year Plan. Some units have two balconies – one at each end. Adjacent bays can be combined, allowing the Support to accommodate units ranging in size from 51m² to 117m².

　　Placing the vertical pipe-shaft at the head of the common stairway places all bathrooms and kitchens along the party wall between units. Nonetheless, the size, configuration and exact location of each of these spaces is variable, enabling a useful variety of unit layouts. Toilets are located close to the vertical drain stack. Placing horizontal piping behind kitchen base cabinets allows kitchens to be placed in a variety of locations.

　　The design emphasizes the possibility of better variability in kitchen and bathroom planning and location, as a means to create greater flexibility within housing units. A second goal was to rationalize the installation and demounting of partitions and other interior systems. The systems used present three major improvements:

1. Placing a vertical pipe chase at each public stair. Centrally locating this base building infrastructure element allows variation in the position of the kitchen and bathroom in each unit.
2. Improving Support/Infill coordination, particularly in regard to partitions and horizontal piping.
3. Offering five partitioning systems, from foreign companies including US Gypsum as well as from local Chinese companies (including a shipbuilder). Showcasing variety was also a means of promoting the industrialization of housing infill in China.

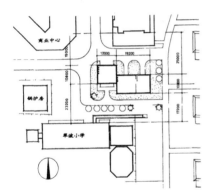

Fig. 4.89 Site plan. Drawing courtesy of M & A Architects.

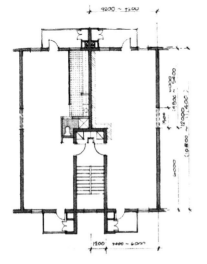

Fig. 4.90 Support plan. Drawing by Stephen Kendall, after original by M & A Architects.

Fig. 4.91 Building with dwelling layouts. Drawing courtesy of M & A Architects.

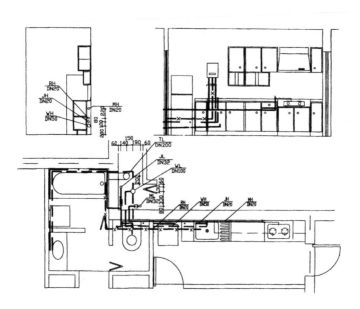

Fig. 4.92 Technical diagram of piping. Drawing courtesy of M & A Architects.

1995   VVO/Laivalahdenkaari 18
Helsinki, Finland

Fig. 4.97 Photograph courtesy of Jussi Tiainen.

ARCHITECT: Arkkitehtuuri Oy Kahri & Co.
OWNER: VVO Rakennuttajat Oy
DWELLINGS: 97 rental units
SUPPORT CONSTRUCTION: [See below]
INFILL PROVISION: [See below]

This 9092m² project, an apartment block of 5–6 stories, incorporates into one project the most comprehensive developments toward Open Building in Finland. Highly adaptable technology was utilized in connection with a successful user participation procedure and financing via government loans.

Building systems used in the project include:

- reinforced concrete element frame with hollow core slabs and load bearing walls between some apartments (in some cases, walls between units allow small units to be combined in future)
- supply services independently distributed to each dwelling
- separate intake/exhaust ventilation with heat recovery in each dwelling
- radiator-free floor heating
- prefabricated box-unit balconies with optional materials, finishes and fittings
- demountable partitions for future reconfiguration

- fixed bathrooms, with the remainder of the floor plan, including kitchens, freely divisible by user.

The user participation process was equally developed. Future residents met as a group on a number of occasions prior to implementation. Separate meetings with the architect were also held for each household, out of which the architect developed up to six optional floor plans for the individual client. Each option included the price of various choices, including finishes, balcony railings, fenestration and so on.

With the exception of the fixed bathroom spaces, households could decide on their own dwelling unit floor plan. In 70% of the units, occupants selected the floor plans, finishes and equipment. In addition, residents were consulted regarding the location of their apartment within the building, the fenestration, and the balcony's balustrade design. Households also participated in the Support design process.

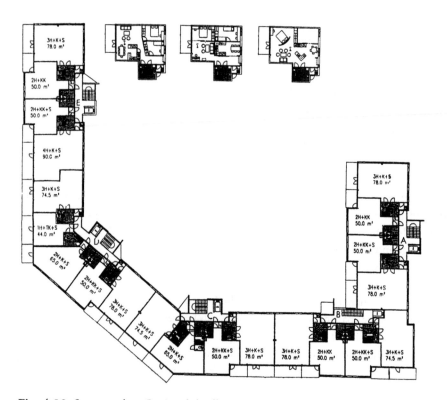

Fig. 4.98 Support plan. Optional dwelling unit plans are shown above. Drawing courtesy of Architecture Office Kahri and Co.

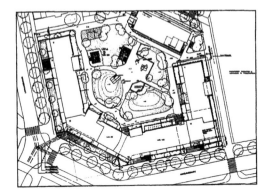

Fig. 4.99 Site plan. Drawing courtesy of Architecture Office Kahri and Co.

Fig. 4.100 Optional dwelling unit plans. Drawing courtesy of Architecture Office Kahri and Co.

1996   Gespleten Hendrik Noord
Amsterdam, Netherlands

Fig. 4.93 Photograph courtesy of Luuk Kramer.

ARCHITECT: De Jager & Lette Architecten, Van Seumeren, Van der Werf
INITIATOR: Stichting Medio Mokum and Woonstichting De Key
DWELLINGS: 28
SUPPORT CONSTRUCTION: Skip-floor corridors; concrete slab and party walls
INFILL PROVISION: User-selected layouts using off-the-shelf subsystems

Within older sections of Amsterdam there is often no possibility to move to better accommodations. This situation led five families to create an alternative solution. They selected a site that offered space for 28 apartments. Building for themselves allowed the participants to save money, which could then be invested in extra quality, in adapting the apartments to the specific requirements of the households involved and in added attention to the architecture. The building has 16 government-subsidized apartments, average price Dfl. 188 000 and 12 'free sector' apartments, average price Dfl. 214 000.

The design process was divided into two stages. Prior to choosing their own apartments, prospective buyers discussed the complex as a whole: its functions, the number of units, the price level, common spaces, the facade, standard unit layouts and priorities within the building budget. After agreement was reached about these issues they began a second stage, in which apartments were assigned and individual requirements discussed.

The design of the complex features a gradual transition from public to private. A central entrance with internal corridors gives access to the building. On the ground floor is a large hall with one entrance from the courtyard and one from the street. In addition to an elegant flight of stairs leading up to the internal corridor on the second floor, there is also an elevator. The fifth floor corridor is accessed by elevator or staircase.

The apartments are tuned to the individual wishes of the participants, and are larger and more spacious than is usual in this price range. In addition to sharing decision-making in the project, occupants share the large central hall on the first and second floors and an inner garden, which provides a protected playground for children. Almost all apartments incorporate a small mezzanine in 1-1/2 stories and have a flexible floor plan; only the party walls between units are Support walls. Whenever possible, main utility shafts are situated at the center of the apartments. This enables the bathroom, kitchen and toilet to be grouped against one of the two Support walls, or placed in the middle of the apartment. Close consultation between the occupants, the architect and the contractor eventually resulted in 28 different floor plans. Layout and function of the various spaces are easily adaptable, offering an almost unlimited diversity in apartment types and sizes, floor plans and finishing, as well as a possibility of other complementary uses. The complex also has a high long-term value. Despite this investment in capacity, construction of the apartments remained highly economical.

In architectural terms, the housing complex has been conceived both as an independent entity and as part of the city. Tall brickwork towers alternate with lower horizontal facades in wood. The towers refer to the brickwork corner sections of adjacent buildings. The facade subtly reflects the diversity of the apartments behind it.

Partly owing to the 'open subscription principle' – a process of bringing new people into the condominium – a healthy social bond has grown between occupants. They respect one another's privacy and at the same time seek to maintain their mutual commitment.

Fig. 4.94 View of the inner courtyard. Photograph courtesy of Joanne de Jager.

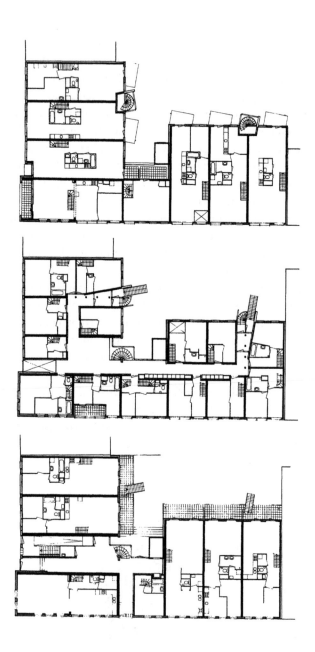

Fig. 4.95 Floor plans of dwelling units.
Drawing courtesy of De Jager & Lette Architects.

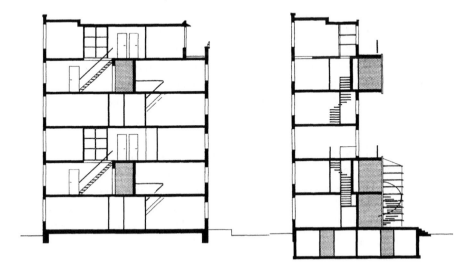

Fig. 4.96 Building cross-sections showing corridors on alternating floors. Drawing courtesy of De Jager & Lette Architects.

1996– Tsukuba Two Step Housing
Tsukuba, Japan

Fig. 4.101 Photograph courtesy of Building Research Institute, Ministry of Construction.

ARCHITECT: Building Research Institute/Ministry of Construction
SPONSOR: Building Research Institute/MOC/Dweller's Cooperative
DWELLINGS: Project #1(15 units); Project #2 (4); Project #3 (11); Project #4 (13);
    Project #5 (12); Project #6 (12); Project #7 (10); Project #8 (10).
SUPPORT CONSTRUCTION: Reinforced concrete; some projects use 'upside-down'
    slab/beam construction; some use ordinary slab and flat beam.
INFILL PROVISION: Most use conventional interior construction; some use new
    products; some use raised floors beneath conventional infill.

The conventional practice of housing developers in Japan has traditionally precluded household involvement in determining dwelling character or quality. Couples start out in 'standard quality' multi-family units, live there for 5–10 years, then move to a single family residence until old age. At that time, they move again: to an apartment. Within that traditional cycle of housing, however, the eagerly anticipated stage of buying a house is now almost impossible to achieve. This results in part from Japan's unique laws surrounding land tenure.

The Tsukuba Method serves as a demonstration of a new way to finance and build housing. It represents an application and an extension of the Two Step Housing System.

It aims to demonstrate that it can be worthwhile for families to remain in their dwelling within a multi-family building, both financially and in terms of comfort. To accomplish that demonstration requires adjusting the ownership and financial structure of housing, as well as the method of building. Therefore, one main goal of this housing initiative is to implement a new form of land ownership: Normally in Japan, a landowner will deed 'right of use' of land to a second party, who may subsequently construct a building. The landowner retains ownership, and can, in theory, reclaim right of use. In reality, however, the landowner can never successfully evict either any party granted the right of use, or their successors. Therefore, the land owner cannot truly regain possession. This reduces the rationale for selling in the first place. Although high costs frequently prevent landowners themselves from developing land, landowner incentives to sell right of use are few. As a consequence, it is difficult and costly to find land on which to build housing.

In the Tsukuba Method, the landowner in essence leases land to a Cooperative Association, while retaining title to it. In exchange for severely limiting their right of use, coop members enjoy lowered initial costs and predictable long-term costs. Coop members own their units for the first 30 years. In the 31st year, title to the land reverts to the landlord, who, by prior contract, begins renting to the households. For the next 30 years, occupants pay only a repair/maintenance fee, as in a condominium, plus a small monthly assessment for renting the land. At year 60, all units are automatically leased, at the market rate. To help solve financing problems, the Japanese Government Housing Loan Corporation developed a new leasehold loan. This unique loan places the mortgage only on the building, not on the land. As a result, it is now possible to borrow 80% of the price of the housing without mortgaging the land. This is an important innovation: in the Japanese system of real property, the ownership of land and building can be divided and the mortgage value of the land is higher than that of the building. In this sense, GHLC is a co-inventor of the Tsukuba Method.

A number of such projects have been built on a deliberately small scale. The first, with 15 dwellings, was built in Tsukuba in 1996. A second project of four units was built there in 1997. A third project, in Tokyo, was built with 11 dwellings. The first three were experimental projects led by the Building Research Institute team. Five additional projects have been built in the private sector. Project #6 is a pioneering project in another sense: in Japan, construction is legally finished and a building ready to be occupied only after all units are finished. This has posed a legal barrier for the Two Step Housing Method. Project #6, the Turumi Project in Yokohama, will be the first case in Japan in which dwellers can occupy units before all are finished.

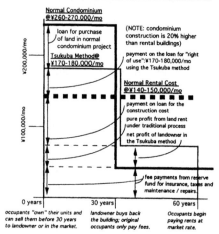

Cost Comparison of Three Occupancy Types:
Rental, Condominium and the Tsukuba Method

Fig. 4.102 Comparative diagram of financing methods. Drawing by Stephen Kendall, after an original by Hideki Kobayashi, Building Research Institute, Ministry of Construction.

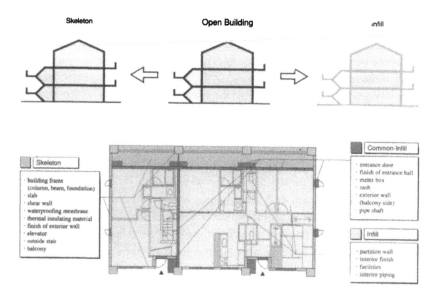

Fig. 4.103 Diagrams of the Two Step Supply System concept and a floor plan from one project. Drawings courtesy of Building Research Institute, Ministry of Construction.

1997 Hyogo Century Housing Project
Hyogo Prefecture, Japan

Fig. 4.104 Photograph courtesy of Seiichi Fukao.

ARCHITECT: Hyogo Prefecture Housing Authority + Ichiura Architects
OWNER: Hyogo Prefecture Housing Authority
DWELLINGS: 104 dwelling units
SUPPORT CONSTRUCTION: 'Upside-down' slab/beam floor system
INFILL PROVISION: Raised floor, unit bath

This project, developed and owned by the Hyogo Prefecture Housing Authority, is a rental project combining the principles of the Two Step Housing Supply and Century Housing Systems. The project contains 48 units of 80m², 36 units of 92m², and 12 large 124m² units. One parking space per dwelling unit is available on-site. A community room was built as part of the project. Community garden plots are made available at various places in and around the site.

The project focussed on three principle objectives:

- long life for the common elements
- flexibility for short-term elements
- accessibility for elderly and handicapped users and visitors

The skeleton and infill are designed as distinct technical systems. The skeleton is designed for 100-year durability and also to accommodate the expected changes in unit size and layout over the years. It uses the principle of the 'upside-down' slab-and-beam, producing a flat ceiling for the unit below, and a pipe chase under a secondary, raised floor for the unit above. Pipe sleeves allow passage of horizontal drain and supply lines, ducts and cables through the up-turned beams.

A second set of components constitutes the exterior walls, facades and party walls between dwelling units. This set of components also includes the roof and the main piping, cabling and mechanical equipment risers that supply utility closets outside the front door of each dwelling unit.

A third set of components includes the interior infill for each individual unit. This consists of partitions (both fixed and movable), doors and door frames and all interior finishes. It also includes cabinets, storage units and unit-specific plumbing, wiring, heating and air conditioning equipment. The raised floor in conjunction with the upside-down floor structure provides an opportunity to build large storage units accessed by floor hatches, for example in the kitchen.

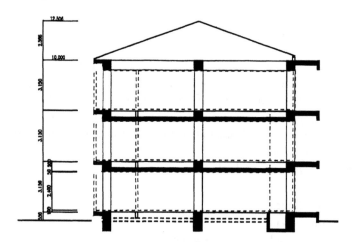

Fig. 4.105 Schematic cross-section showing 'upside-down' slab/beam. Drawing courtesy of Hyogo Prefecture Housing Authority.

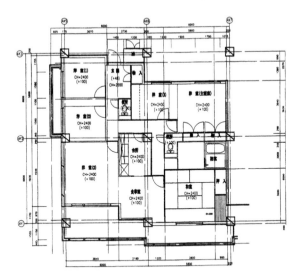

Fig. 4.106 Dwelling unit plan. Drawing courtesy of Ichiura Architects.

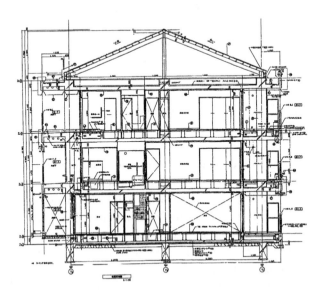

Fig. 4.107 Construction drawing section showing interior elevation of infill.
Drawing courtesy of Ichiura Architects.

1998– Yoshida Next Generation Housing Project
Osaka, Japan

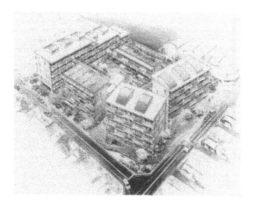

Fig. 4.108 Drawing courtesy of Kentiku Kankyo Kenkyujo.

ARCHITECT: Kenchiku Kankyo Kenkyujo and Shu-Koh-Sha Architecture and Urban
     Design Studio
OWNER: Osaka Prefecture Housing Supply Corporation
PLANNER: Construction Committee of Next Generation Urban Housing + Tatsumi
     and Takada
DWELLINGS: 53 dwelling units
SUPPORT CONSTRUCTION: 'Upside-down' slab/beam floor system
INFILL PROVISION: Matsushita Electric, Daiken, Panekyo

Planning for this project started in 1995. Previous Two Step projects had offered units
for sale. Now the aim was to adapt the system for the rental market. Ultimately, the
Yoshida project defines a third 'hybrid' way of supplying housing: Infill is neither entirely owned nor entirely rented. The concept advanced by Professors Tatsumi and Takada
of Kyoto University allows occupants to own a limited part of the total dwelling infill:
for instance, only demountable or movable walls and furniture. Storage units
(wardrobes, closets, etc.) are not owned by the inhabitants but form part of the Support:
they are rented. Furthermore, by prior agreement, maintenance of the infill owned by
dwellers is provided by a group of infill suppliers. A standard plan fixes the location of
sanitary unit, kitchen and any fixed storage elements. These are not affected by tenant
preferences.

The Osaka Prefecture Housing Supply Corporation intends to supply only the project skeleton in future. Private industry is expected to supply all additional elements as infill. The Corporation's intention is also to refurbish existing rental housing units using infill systems. To that end, the Corporation has directed participating companies to supply infill systems whose primary components, including partitions, doors and cabinets, are: easily installed by do-it-yourself (DIY) or unskilled workers; owned by dwellers; and inexpensive to purchase and install. In addition, these components must not include wiring, plumbing or mechanical equipment, nor require any cutting beyond trimming end panels to meet specific unit conditions. In addition, the infill components need not adhere to any acoustical or thermal performance specifications.

The skeleton uses an inverted slab/beam floor system similar to that in the Hyogo Century Housing Project and some of the Tsukuba Two-Step Housing projects.

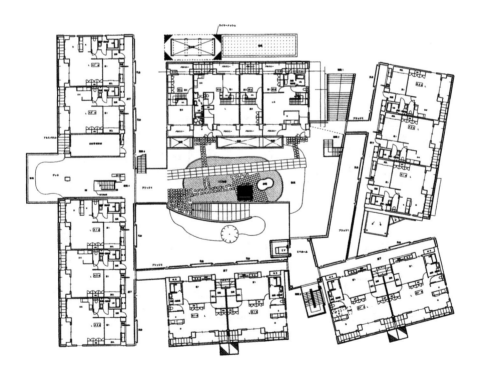

Fig. 4.109 Plaza level building and site plan. Drawing courtesy of Kentiku Kankyo Kenkyujo.

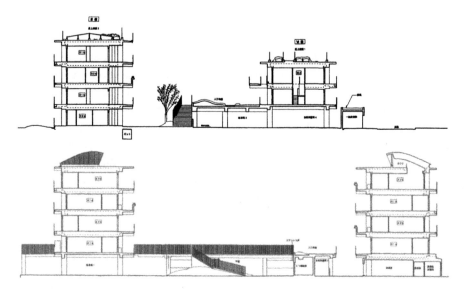

Fig. 4.110 Site and building cross-section views. Drawing courtesy of Kentiku Kankyo Kenkyujo.

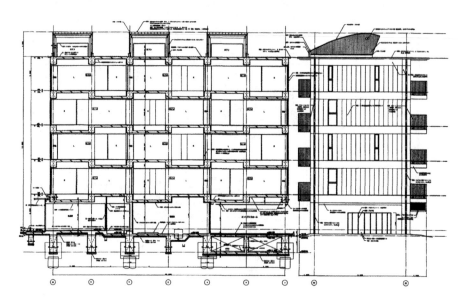

Fig. 4.111 Longitudinal section showing stepped Support floor. Drawing courtesy of Kentiku Kankyo Kenkyujo.

1998 The Pelgromhof
Zevenaar, Netherlands

Fig. 4.112 Rendering by Ingolf Kruseman. Courtesy of ASK.

ARCHITECT: Frans van der Werf
OWNER: Algemene Stichting Woningbouw Zevenaar and Pelgromstiching, Zevenaar
DWELLINGS: 215
SUPPORT CONSTRUCTION: Concrete slab and party walls
INFILL PROVISION: User-selected layouts using off-the-shelf subsystems

This project is funded by the Zevenaar Residential Construction Foundation and the Pelgrom Foundation. In a newly-constructed project of 215 dwellings for the elderly, it combines principles of Open Building, ecological/sustainable design and organic architecture. The project incorporates 169 apartments for independent living, parking for 86 cars, and a nucleus of 46 units designed for assisted or intermural care living. The project also has a reception room, a social center with kitchen, a restaurant, theater, shop and library. Project cost was in excess of Dfl. 50 million (US $25 million). The project was awarded experimental status by the Dutch government and was also selected as a National Model of Sustainable and Energy-efficient Construction by the Ministry of Housing. The Pelgromhof meets criteria of the Dutch National Sustainable Construction Measures for Residential Building, but goes further by using natural paint and heat pumps.

Architect Frans van der Werf has been realizing Support/Infill projects for more than two decades. This recent work embodies many fundamental principles of Open Building:

- *Open construction:* Each resident has a hand in creating a place which corresponds to his or her own way of life. Occupants lay out their own dwellings using a full-scale mock-up model.
- *Life-time guaranteed dwelling:* The project offers living space for households in different later stages of life, popularly known as 'go-go's, slow-go's and no-go's.' It answers senior needs for accessibility, safety and adaptability.
- *Social cohesion:* Full social integration of older people who will require some assistance. The Pelgromhof provides tailored care and a safe, tranquil yet vital environment located in the city center.
- *Organic architecture:* The project's shapes and colors and landscaping, in keeping with the owners' philosophy, are intended to house residents in communion with nature. The site features abundant plantings in planting beds and on external walls and roofs, in addition to air purification, specimen trees, flowing water and a diversity of flora and fauna.
- *Digital superhighway:* telemetering to aid safety, communications and energy management.
- *Sustainable construction:* Among the many green architecture features of the building are: bio-ecological paints and other materials; new high-efficiency floor heating; reduced use of concrete; heating with solar energy; application of individual and collective heat pumps for energy savings in climate control; and optimization of window and roof insulation.

The target group for the project consists of residential consumers in the 50+ age group who want lifetime guaranteed dwellings – dwellings which remain suitable throughout different stages of life. The opportunity for individual choice of layout has never before been implemented on such a large scale in the rental sector.

The Pelgromhof was recently nominated for the Dutch Building Award. A second phase was under construction as of this writing.

Fig. 4.113 Aerial view of the project in the neighborhood. Photomontage by Michel Hofmeester. Courtesy of AeroCamera BV.

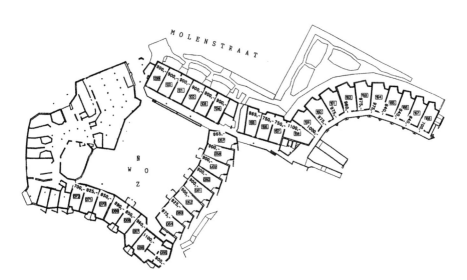

Fig. 4.114 Support plan at ground level. Drawing courtesy of Frans van der Werf.

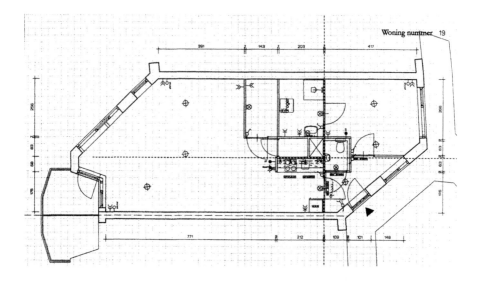

Fig. 4.115　Dwelling plan variant. Drawing courtesy of Frans van der Werf.

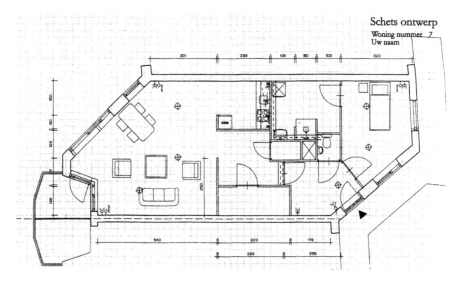

Fig. 4.116　Dwelling plan variant. Drawing courtesy of Frans van der Werf.

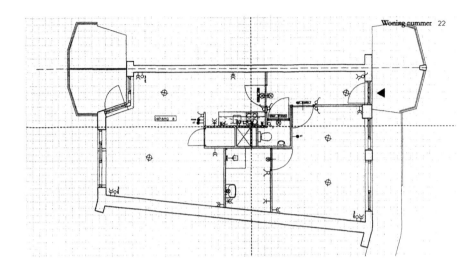

Fig. 4.117 Dwelling plan variant. Drawing courtesy of Frans van der Werf.

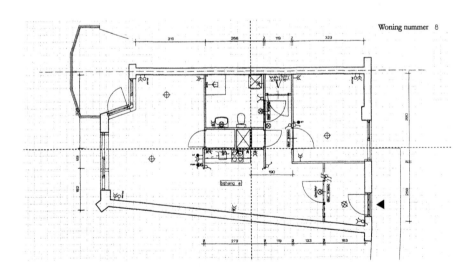

Fig. 4.118 Dwelling plan variant. Drawing courtesy of Frans van der Werf.

1998   HUDc KSI 98 Demonstration Project
Hachioji, Japan

Fig. 4.119 View of the Z-beam at the entry of a dwelling unit.
Photograph courtesy of HUDc Research Laboratory.

ARCHITECT: HUDc Design Office and Kan Sogo Design Office
OWNER: Housing and Urban Development corporation
DWELLINGS: 5 experimental units and 2 penthouse units
SUPPORT CONSTRUCTION: Concrete 'Z-Beam' skeleton; concrete hollow flat slab,
    beams and columns using post-stressed pre-cast concrete; common drainage lines
    set outside of each dwelling unit
INFILL PROVISION: HUDc Infill and a variety of private sector infill products.

The Housing and Urban Development corporation (HUDc) has for many years been a
major public sector developer of new town and condominium projects. It also owns
more than 720 000 rental housing units. As that housing stock ages, the corporation's
main focus is shifting to inner city regeneration and the provision of rental housing.
Experimental projects such as KSI are a direct result of this changing mandate.

In Japan, the demand for multi-unit consumer-oriented rental housing remains
unmet. Condominiums are not a preferred option: most do not easily adapt in response
to inhabitant preferences. A very large percentage of occupants of condominiums must
agree to any or all upgrades or repair work. There are also heavy restrictions on inhabitant remodeling, a heavy burden of restoration costs after earthquakes, and so on.

However, in the housing system proposed in the KSI project, the public sector owns the Support and individual residents rent space and own the infill. HUDc hopes to develop new S/I housing technologies and to disseminate S/I housing throughout Japan. In order to succeed, new infill systems are needed for both new construction and renovations.

Over the years, HUDc had developed four Support structural systems to provide the capacity needed for S/I housing. These include: partially reversed beam structure; shear wall structure with partially sunken slab; flat beam structure; and a solid frame structure with no shear or bearing walls. To these systems, the KSI 98 project now adds a fifth, a Z-beam structure. In the KSI 98 Demonstration Project, the first floor has an exhibition space and two model units for private sector infill demonstrations. The second floor has the HUDc infill system model home plus two more units modelling private sector infill. The third floor contains two penthouses which utilize new proprietary technologies. Throughout construction, a number of developments are to be monitored, including: subsystem interfaces; new product performance; and experimental techniques to shorten construction time and minimize the number of on-site workers representing different trades.

In the HUDc Infill System, low-slope drain lines (modeled after the Matura Infill System's zero-slope piping), vinyl sheathed wiring, and 'manifold' water supply pipes are installed above the concrete floor. A raised floor supported by adjustable pedestals with rubber tips, similar to a low-cost computer access floor, is laid down as a sub-floor. The raised floor and the partitions with gypsum board attached to one side are pre-assembled in a factory. Electrical wiring is put in place in the walls, as well as in a recessed wiring 'trench' at the perimeter of rooms in the zone of the raised floor. Flat cables provided by Matsushita Electric Works and other companies are used on the ceiling. Independent home-run lines extend from each fixture to a header located outside the front door. These connect under a raised floor to the vertical stack. Supply pipes are also home-run from manifolds, one for hot, the other for cold water.

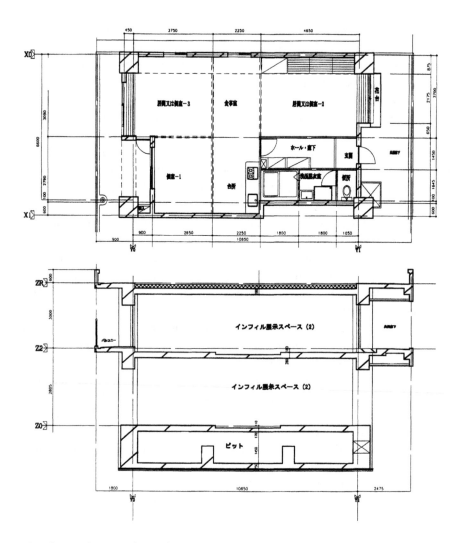

Fig. 4.120 Support plan and section. Drawing courtesy of HUDc Research Laboratory.

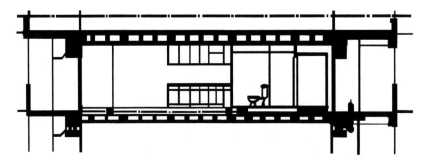

Fig. 4.121 Cross-section of a dwelling unit. Drawing by Jennifer Wrobleski, after original by HUDc Research Laboratory.

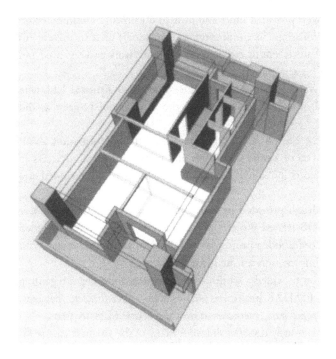

Fig. 1.122 Schematic view of the Support. Drawing courtesy of HUDc Research Laboratory.

ACKNOWLEDGMENTS

Case Studies

Many colleagues from around the world have directly and substantively contributed materials to add to the case studies presented here. Others have read and corrected text, offered suggestions or introduced us to OB projects and their creators.

Netherlands: Ype Cuperus continually contributed to the case studies' substance and breadth of coverage, providing and correcting information, and facilitating contacts throughout the OB network. Karel Dekker contributed information about current renovation work in the Netherlands. John Habraken offered encouragement, analysis, comment and correction. He reviewed an early draft, apprised us of new OB developments and also introduced us to several architects and projects. Joanne de Jager provided numerous images of her project and helped craft the text for it. Fokke de Jong searched through his archives for photographs of early OB projects and provided additional information and advice. Henk Reijenga sent information about his work at the tissue level. Frans van der Werf provided slides and printed documents, commented on drafts and arranged for permissions in connection with materials used to present his pioneering projects. Bert Wauben sent us excellent images of his work and reviewed and corrected the case study of his seminal project.

Austria: Architect Jos Weber generously put us in contact with Ottokar Uhl, who provided photographs and other documentation of the Hollabrunn project, as did Architect Franz Kuzmich.

Belgium: The Office of Lucien Kroll provided a wealth of information about Kroll's work, including several of Kroll's CD-Roms and books and ongoing access to superb images of many projects including 'La Mémé.' Kroll also kindly reviewed and corrected a draft case study.

Switzerland: Marcus Heggli provided us with copies of Henz (1995) *Anpassbare Wohnungen,* and a list of OB-related projects and architects. Among those included were Martin Erni, whose Davidsboden project is the subject of a case study; and Willi Rusterholz, who provided information for the Neuwil case study.

UK: Nicholas Wilkinson provided a wealth of information and images regarding the Wilkinson and Hamdi PSSHAK project, as well as many other projects that have appeared in the pages of *Open House International* over the course of many years.

Japan: Seiichi Fukao generously donated slides for many of the Japanese case studies. He graciously reviewed the case studies and corrected many parts of the text, providing much crucial background information. Shinichi Chikazumi read with similar care many parts of the text. His background commentary, corrections and improvements are also reflected in many case studies. Mitsuo Takada offered timely thoughts and wording for a number of projects in which he has been involved. Kazuo Kamata and Seiji Kamata of HUDc provided invaluable information and documentation on the long list of HUDc OB-related projects and research activities. Midori Kamo, Osaka Gas Corporation, helped obtain permission to use materials related to the Next21 project case study. Dr. Hideki Kobayashi provided numerous images and supporting information about the Tsukuba Method.

China: Jia-sheng Bao provided drawings and photographs of his pioneering OB project in Wuxi. Zhang Qinnan provided drawings and photographs of the project he helped bring to fruition in Beijing. Jia Beisi's *Housing Adaptability Design* proved a substantial source book regarding early project developments toward Open Building in Europe as well as in China.

United States: Koryn Rolstad assisted in obtaining photographs and drawings of the Banner Building project and in obtaining permissions for their use.

Finland: Esko Kahri provided drawings and photographs of his OB project and Ulpu Tiuri consistently provided a great deal of insight and information, and generously provided background on a number of early Swedish and German OB projects, based on her own research.

Among the recommended list of Additional Readings books and journals that follows, we are indebted to several as primary resources for many case studies presented. Among those, we have frequently consulted *Scope of Social Architecture,* the fine book edited by C. Richard Hatch (1976); (1976) *Industrialization Forum (IF)* **7** no. 1; and (1987) 'Changing Patterns in Japanese Housing,' a special issue of *Open House International* **12** no. 2. *Open House International* remains the foremost journal devoted to examining the breadth and depth of OB internationally. Many of the early European projects were first published in *Plan.* Much information in the Japanese OB case studies originally appeared in English in *Developments Toward Open Building in Japan* (Kendall, 1997).

ADDITIONAL READINGS

1966 Neuwil

Metron Architects. (1966) Überbauung 'Neuwil' in Wohlen AG. *Werk.* February.

Henz, A. and Henz, H. (1995) Anpassbare Wohnungen. *ETH Wohnforum.* TH Hönggerberg, Zurich.

Beisi, J. (1994) *Housing Adaptability Design.* ETH Zurich, Post-doctoral Thesis, Zürich.

1974 Maison Médicale (La Mémé), Catholic University of Louvain

Froyen, H-P. (1976) Structures and Infills in Practice – Four Recent Projects. *Industrialization Forum.* **7** no. 1. pp. 17–19.

Kroll, L. (1984) Anarchitecture, in *The Scope of Social Architecture.* (ed C.R. Hatch) Van Nostrand, New York.

Kroll, L. (1985) CAD-Architekture, in *Vielfalt durch Partizipation, Vorwort von Ottokar Uhl.* Verlag C.F. Müller, Karlsruhe.

Kroll, L. (1987) *An Architecture of Complexity*, MIT Press, Cambridge.

Kroll, L. (1987) *Buildings And Projects*, Rizzoli, New York.

Kroll, L. (1996) Bio, Psycho, Socio/Eco in *Ecologies Urbaines.* (preface by ed Pierre Loze) L'Harmattan, France.

Besch, D. (1996) *Werken van het Atelier Lucien Kroll*, Delft Univesity Press, Netherlands.

1976 Dwelling of Tomorrow

Dirisamer, R., Kuzmich, F., Uhl, O., Voss, W., Weber, J.P. (1976) Project Dwelling of Tomorrow, Hollabrunn, Austria, *Industrialization Forum.* **7** no. 1. pp. 11–16.

Dirisamer, R., Dulosy, E., Gschnitzer, R., Kuzmich, F., Panzhauser, E., Uhl, O., Voss, W., Weber, J. (1978) *Forschungsbericht 1: Wohnen Morgen Hollabrunn.* Arbeitsgemeinshaft fur Architektur, Vienna.

Uhl, O. (1984) Democracy in Housing, in *The Scope of Social Architecture*, (ed C.R. Hatch). Van Nostrand, New York. pp. 40–47.

1977 Beverwaard Urban District

Carp, J. (1979) SAR Tissue Method: An Aid for Producers, *Open House.* **4** no. 2. pp. 2–7.

Reijenga, H. (1981) Town Planning Without Frills. *Open House.* **6** no. 4. pp. 10–20.

Reijenga, H. (1977) Beverwaard. *Open House.* **2** no. 4. pp. 2–9.

1977 Sterrenburg III

De Jong, F.M. (1979) Sterrenburg III, Dordrecht: Support/Infill Housing Project. *Open House.* **4** no. 3. pp. 5–21.

1977 Papendrecht

Van Rooij, T. (1978) Molenvliet: Support Housing for the rented sector recently completed in Papendrecht, Holland. *Open House.* **3** no. 2. pp. 2–11.

van der Werf, F. (1980) Molenvliet-Wilgendonk: Experimental Housing Project, Papendrecht, The Netherlands. *The Harvard Architecture Review: Beyond the Modern Movement.* **1** Spring.

van der Werf, F. (1984) A Vital Balance, in *The Scope of Social Architecture.* (ed C.R. Hatch). Van Nostrand, New York. pp. 29–35.

van der Werf, F. (1993) *Open Ontwerpen.* Uitgeverij 101, Rotterdam.

1979 PSSHAK/Adelaide Road

Hamdi, N. (1978) PSSHAK, Adelaide Road, London. *Open House.* **3** no. 2. pp. 18–42.

Hamdi, N. (1984) PSSHAK: Primary Support Structures and Housing Assembly Kits, in *The Scope of Social Architecture.* (ed C.R. Hatch). Van Nostrand, New York. pp. 48–60.

Hamdi, N. (1991) *Housing without Houses: Participation, Flexibility and Enablement.* Van Nostrand, New York.

1979 Hasselderveld

Wauben, B. (1980) Experimental Housing, Haeselderveld, Geleen, Holland. *Open House.* **5** no. 3. pp. 11–17.

Wauben, B. (1985) Experimental Housing, Haeselderveld, Geleen, Holland. *Bouw* **4**. 2/16.

1983 Estate Tsurumaki /Town Estate Tsurumaki

Fukao, S. (1987) Century Housing System: Background and Status Report. Changing Patterns In Japanese Housing (ed S. Kendall). Special issue, *Open House International.* **12** no. 2. pp. 30–37.

1984 Keyenburg

Monroy, M.R. and Geraedts, R.P. (1983) May we add another wall, Mrs. Jones?, *Open House International.* **8** no. 3. pp. 3–9.

Norsa, A. (1984) E l'Olanda batte il Belgio, il Successo di Keyenburg, *Construire.* no. 21 Luglio/Agosto.

Carp, J. (1985) *Keyenburg: A Pilot Project.* Stichting Architecten Research, Eindhoven.

1985 Free Plan Rental

Fukao, S. (1987) Century Housing System: Background and Status Report. Changing Patterns In Japanese Housing (ed S. Kendall). Special issue, *Open House International.* **12** no. 2. pp. 30–37.

1987 Support Housing , Wuxi

Bao, J-S. (1987) Support Housing in Wuxi Jiangsu: User Interventions in the Peoples' Republic of China. Changing Patterns In Japanese Housing (ed S. Kendall). Special issue, *Open House International.* **12** no. 1. pp. 7–19.

1989 Senri Inokodani

Tatsumi, K. and Takada, M. (1987) Two Step Housing System. Changing Patterns In Japanese Housing (ed. S. Kendall). Special issue, *Open House International.* **12** no. 2. pp. 20–29.

Kendall, S. (1995) *Developments Toward Open Building In Japan.* Silver Spring, Maryland. pp. 10–11.

1990 Patrimoniums Woningen

Yagi, K. (ed) (1993) Renovation by Open Building System. *Process Architecture.* no. 112.

Dekker, K. (1998) Consumer Oriented Renovation of Apartments – Voorburg, the Netherlands, *CIB Best Practices Papers.* CIB Web Site (www.cibworld.nl).

Cuperus, Y., and Kapteijns, J. (1993) Open Building Strategies in Post War Housing Estates, *Open House International.* **18** no. 2. pp. 3–14.

1991 'Davidsboden' Apartments

Christoph Merian Stiftung. (1992), *Wohnsiedlung Davidsboden Basel. Ein Neues Wohnbodell der Christoph Merian Stiftung.* Christoph Merian Stiftung, Basel.

Beisi, J. (1994) *Housing Adaptability Design.* ETH Zurich Post-doctoral Thesis, Zürich.

Henz, A. and Henz, H. (1995) *Anpassbare Wohnungen.* ETH Hönggerberg, Zürich.

1993 Green Village Utsugidai

Kendall, S. (1995) *Developments Toward Open Building In Japan.* Silver Spring, Maryland. pp. 12–13.

1994 Banner Building

(1995) Coming: Housing that Looks Like America. *Architectural Record.* January. pp. 84–88.

(1996) AIA Honor Awards. *Architecture.* May.

1994 Next21

Itoh, K. (ed) (1994) Next21. Special issue, *SD 25*.

(1994) Next21. *Kenchiku Bunka*. 567 January.

1994 Pipe-Stairwell Adaptable Housing

Ma Y., Zhang Q. and Research Team on Universal Infill System in Adaptive Housing (1995) *Design Collection for the XiaoKang Type Flexible Space Housing*. Beijing, China.

Research Team on Adaptive Housing / M & A Architects and Consultants International. (1995) Modular Coordination in Housing Design. *Architectural Journal*. May.

Research Team on Adaptive Housing / M & A Architects and Consultants International. (1995) Service System Design in Adaptive Housing. *Architectural Journal*. September.

1996 Gespleten Hendrik Noord

de Jager, J. *et al.* (1997) Gespleten Hendrik Noord in Amsterdam. *Bouw*. March.

de Jager, J. *et al.* (1997) New Housing For Families in Amsterdam – Gespleten Hendrik Noord. *Bouwwereld*. June 16.

de Jager, J. (1998) Gespleten Hendrik Noord. *Westerpark: Architecture in a Dutch City Quarter, 1990–1998*. NAi Press, Amsterdam.

1995 VVO/Laivalahdenkaari

Kautto, J., Kulovesi, J., Pekkanen, J., Tiuri, U., (ed P. Huovila). (1998) *Milieu 2000: Experimental Urban Housing: Four Pilot Projects in Helsinki, Finland. City of Helsinki*. TEKES, Ministry of the Environment, Helsinki.

Tiuri, U. and Hedman, M. (1998) *Developments Towards Open Building In Finland*. Helsinki University of Technology, Department of Architecture, Helsinki.

Tiuri, U. (1998) Open Building – Housing for Real People. Arkkitehti 3. pp. 18–23.

1998 The Pelgromhof

van der Werf, F. (1997) Pelgromhof and Open Building. *Gezond Bouwen & Wonen*. 5 Sept/Oct.

van der Werf, F. (1998) Interview with Argo Oskam and Koos Timmermans. *Bouw*. February.

van der Werf, F. (1998) Open Building, Occupant Participation. *Woningraad Magazine*. March.

van der Werf, F. (1998) Open Building, Occupant Participation, *Renovatie & Onderhoud*. May.

1998 Tsukuba Method

Kendall, S. (1995) *Developments Toward Open Building In Japan*. Silver Spring, Maryland. pp. 24–28.

Kobayashi, H. (1997) *The Era of New Housing*. NHK Publishing Co, Tokyo.

Kobayashi, H. (1997) Tsukuba Method – Open Building Supplied by Leasehold. *Housing*. Japan Housing Association.

Kobayashi, H. (1996) Tsukuba Method. *Nikkei Architecture*. **565**.

Kobayashi, H. (1997) Tsukuba Method. *Nikkei Architecture*. **574**.

Kobayashi, H. (1998) Tsukuba Method. *Axis*. **75**.

Kobayashi, H. (1998) Tsukuba Method. *Data Files Of Architectural Design & Detail*. **68**.

1998 HUDc KSI 98 Project

(1998) KSI Experimental Project. *Nikkei Architecture*. October 19.

(1998) KSI Housing. *FORE*. (Bulletin of Real Estate Association). November.

(1999) KSI Housing: Hachioji Research Center of HUDc. *Syukan Koukyuojutaku*. January 13.

(1999) KSI Housing Experimental Project. *Kenchiku Gijutsu*. January.

REALIZED OPEN BUILDING
AND RELATED PROJECTS:
A CHRONOLOGY

1903 Skalitzerstrasse 99, Berlin,
Germany
1927 Häuser am Weissenhof,
Stuttgart, Germany
1935 Complex 'De Eendracht,'
Rotterdam, Netherlands
1950 Wohnblock, Göteborg, Sweden
1954 Flexibla Lägenheter, Göteborg,
Sweden
1955 Mäander-Seidlung, Orebro-
Baronbackarna, Sweden
1956 Housing Project, Tianjing,
China
1959 Kallebäckshuset, Göteborg,
Sweden
1960 Apartment Block, Göteborg,
Sweden
1966 Überbauung Neuwil, Wohlen,
Switzerland
1966 Diset Project, Uppsala, Sweden
1967 Housing Project, Kalmar,
Sweden
1967 Orminge, Stockholm, Sweden
1968 Saalwohnungen, Vienna, Austria
1969 Housing Complex, Horn,
Netherlands
1970 Six Experimental Houses,
Deventer, Netherlands
1970 Haus am Opernplatz, Berlin,
Germany
1971 Housing Project, Kalmar,
Sweden

1972 Elementa '72, Bonn, Germany
1972 'Dwelling of Tomorrow,'
Hollabrunn, Austria
1973 MF-Haus, Rotterdam,
Netherlands
1973 Project 'Steilshoop,' Hamburg,
Germany
1973 MF-Hause 'Urbanes Wohnen,'
Hamburg, Germany
1974 Überbauung Döbeligut,
Oftringen, Switzerland
1974 'La Mémé' medical student
housing, Catholic University of
Louvain, Brussels, Belgium
1974 Vlaardingen, Holy-Noord,
Netherlands
1974 Social Housing, Assen-Pittelo,
Netherlands
1975 Social Housing, Stroinkslanden
(Zuid Enschede), Netherlands
1975 Les Marelles, Paris, France
1975 PSSHAK/Stamford Hill,
London, England
1975 Social Housing, Zwijndrecht
(Walburg II), Netherlands
1975 Housing, Kraaijenstein,
Netherlands
1975 Zutphen, Zwanevlot,
Netherlands
1976 Öxnehaga, Husqvarna, Sweden
1977 Sterrenburg III, Dordrecht,
Netherlands
1977 De Lobben, Houten,
Netherlands
1977 Papendrecht, Molenvliet,
Netherlands
1979 Feilnerpassage Haus 9, Berlin-
Kreuzberg, Germany

1979	PSSHAK/Adelaide Road, London, England
1979	Hasselderveld, Geleen, Netherlands
1980	KEP Maenocho, Tokyo, Japan
1980	Tissue/Support Project, Leusden Center (Hamershof), Netherlands
1980	Housing Project, Ijsselmonde, Netherlands
1982	Lunetten, Utrecht, Netherlands
1982	KEP 'Estate Tsurumaki,' Tama New Town, Japan
1982	KEP 'Town Estate Tsurumaki,' Tama New Town, Japan
1982	Baanstraat, Schiedam, Netherlands
1982	Dronten Zuid, Netherlands
1982	Niewegein, Netherlands
1982	Senboku Momoyamadai Project Sakai City, Osaka, Japan
1983	Estate Tsurumaki and Town Estate Tsurumaki, Tama New Town, Japan
1983	C I Heights, Machida, Machida-shi, Tokyo, Japan
1984	Pastral Haim Eifuku, Suginami-ku, Tokyo, Japan
1984	Keyenburg, Rotterdam, Netherlands
1984	Cherry Heights Kengun, Tokyo, Japan
1985	PIA Century 21, Kanagawa, Japan
1985	L-City New Urayasu, Chiba, Japan
1985	Tsukuba Sakura Complex, Tsukuba, Japan
1986	'Free Plan Rental Project,' Hikarigaoka, Nerima-ku, Japan
1986	Schauberg Hünenberg, Hünenberg, Switzerland
1986	Terada-machi Housing, Osaka, Japan
1987	Support Housing, Wuxi, China
1987	Tissue Project, Claeverenblad/Wildenburg, Netherlands
1987	MMHK CHS Projects, Chiba, Japan
1987	Yao Minami Housing, Osaka, Japan
1987	Yodogawa Riverside Project #5, Osaka, Japan
1988	Villa Nova Kengun, Kumumoto, Japan
1988	Rune Koiwa Garden House, Tokyo, Japan
1988	Berkenkamp, Enschede, Netherlands
1989	Senri Inokodani Housing Estate Two Step Housing Project, Osaka, Japan
1989	Saison CHS Hamamatsu Model, Shizuoka, Japan
1989	Housing Project, Zestienhovensekade, Rotterdam, Netherlands
1989	Centurion 21, Toyama, Japan
1999	45 three-room-houses, Delft, Netherlands
1990	Hellmutstrasse, Zürich, Switzerland

1990 Support/Infill Project, Eindhoven, Netherlands
1990 Patrimoniums Woningen
Renovation Project, Voorburg,
Netherlands
1990 Herti V, Zug, Switzerland
1990 House #23, Huawei Residential
Quarter, Beijing, China
1990 Residence des Chevreuils, Paris,
France
1991 Hellmutstrasse, Zurich,
Switzerland
1991 'Davidsboden,' Basel,
Switzerland
1991 Flexible Infill Project, Eindhoven, Netherlands
1991 Meerfase-Woningen, Almere,
Netherlands
1991 Schuifdeur-Woning, Amsterdam,
Netherlands
1991 Huawei No. 23, Beijing, China
1992 Patrimoniums Woningen, New
Dwellings, Voorburg,
Netherlands
1992 Experimental House No. 13,
Block 15, Kangjian Residential
Quarter, Shanghai, China
1993 Luzernerring, Basel, Switzerland
1993 Green Village Utsugidai Coop,
Hachioji, Japan
1993– House Japan Project, Tokyo,
Japan
1994 Next21, Osaka, Japan
1994 MIS Project/Shirakibaru Project,
Fukuoka, Japan
1994 42 student apartments, Rotterdam, Netherlands

1994 De Raden Housing Project, Den
Haag, Netherlands
1994 Takenaka Matsuyama Dormitory
Project, Osaka, Japan
1994 Banner Building, Seattle, United
States
1994 Pipe-Stairwell Adaptable
Housing, Cuiwei Residential
Quarter, Beijing, China
1994 Flexible Open Housing with
Elastic Core Zones at Friendship
Road, Tianjin, China
1994 Überbauung 'Im Sydefädeli,'
Zürich, Switzerland
1994 Wohnüberbauung
Wehntalerstrasse-in-Böden,
Zürich, Switzerland
1995 Muracker, Lensburg, Switzerland
1995 Sashigamoi Interior Finishing
Method, Tama New Town,
Tokyo, Japan
1995 Partial Flexible Housing in
Taiyuan, Shanxi Province, China
1995 De Bennekel Housing Project,
Eindhoven, Netherlands
1995 Beiyuan Residential Quarter in
Zhengzhou, Henan Province,
China
1995 Elderly Care Housing, Eijken-
burg, the Hague, Netherlands
1995 53 Houses That Grow, Meppel,
Netherlands
1995 VVO/Laivalahdenkaari 18,
Helsinki, Finland
1995–7 Action Program for Reduction
of Housing Construction Costs,
Hachioji-shi, Tokyo

1996	Block M1-2, Makuhari New Urban Housing District, Chiba, Japan
1996	Tsukuba Method Project #1 (Two Step Housing Supply System), Tsukuba-shi, Ibaraki, Japan
1996	Tsukuba Method Project #2 (Two Step Housing Supply System), Tsukuba-shi, Ibaraki, Japan
1996	Gespleten Hendrik Noord, Amsterdam, Netherlands
1997	Hyogo Century Housing Project, Hyogo Prefecture, Japan
1997	Elsa Tower Project, Tokyo, Japan
1997	HOYA II Project, Tokyo, Japan
1997	6 Support/Infill Houses, Matura Infill, Ureterp, Netherlands
1997	Puntegale Adaptive Reuse Project, Rotterdam, Netherlands
1998	Yoshida Next Generation Housing Project, Osaka, Japan
1998	Sato-Asumisoikeus Oy/ Laivalahdenkaari 9, Helsinki, Finland
1998	Matsubara Apartment/Tsukuba Method Project #3, Tokyo, Japan
1998	Partially Flexible Housing in Beiyuan Residential Quarter, Zhengzhou, Henan Province, China
1998	Housing Tower, Pingdingshan, Henan, China
1998	Essen-Laag, Nieuwerkerk aan de Ijssel, Interlevel infill, Netherlands.
1998	Vrij Entrepot loft residences, Kop van Zuid, Rotterdam, Netherlands
1998	The Pelgromhof, Zevenaar, Gelderland, Netherlands
1998	Support/Infill Project of 8 Houses, Matura Infill, Sleeuwijk, Netherlands
1999	45 three-room-houses in former office, Delft, Netherlands
1999	VZOS Housing Project, the Hague, Netherlands
1999	Tervasviita Apartment Block, Seinäjoki, Finland

PART THREE

METHODS AND PRODUCTS

5

Technical overview

5.1 CHANGES IN NETWORKED RESIDENTIAL BUILDINGS

For over a third of a century, in a variety of ways and settings, movements toward Open Building have paralleled a number of technical developments:

1. *Dwellings are now tethered to multiple networked systems.* Within the space of a century, homes have been transformed. They now incorporate direct links to numerous resource and utility networks: water supply and waste treatment, gas pipelines, electrical power grids, security systems, satellite, cellular and traditional landbased communication networks, television cables and the internet. By contrast, 150 years ago, the dwelling might well have been connected to only the road network, or to none at all.

2. *Almost without exception, these networks now penetrate to the dwelling's core.* Buildings frequently require resource supply outlets serving many complex and interdependent networks throughout every space of the dwelling.

3. *Residential buildings typically lack conduits, raceways, chases or interstitial cavities suited to the distribution of such network parts.* Industry standards, agreements or even precise documentation concerning the placement and interface of network supply lines are not commonly used. Lacking such agreements, supply systems are frequently laid in complete disarray into

spaces within ceilings or walls. There, cables, wires, pipes, ducts and structural elements become hopelessly entertwined.

4. *Within multi-family housing units, collective building structure, building trunk lines and connectors are threaded through dwelling interiors.* Such entanglement of collective infrastructure throughout private space makes it impossible to establish specific boundaries between public and private control and responsibility.

5. *In multi-family buildings, the design of the common structure or skeleton – especially in seismic zones – significantly constrains any attempt to provide reasonably affordable dwelling unit flexibility.* In Japan, as well as in other seismically-sensitive areas, the engineering design of Supports – and of the utility systems that form part of the Support – has strongly influenced developments toward Open Building: Life-safety issues directly constrain the flexibility of spaces.

6. *The gradual accumulation of new networked subsystems in residential buildings over the past 100 years has brought with it an accretion of obsolete construction methods.* In every country where OB is emerging, the obsolescence of rigid trade divisions is becoming clear. The value of multi-skilled installers trained and certified to work across trades is increasingly evident. Similarly, there is an obvious need for products which straddle traditional product/trade classifications. Both infill systems and facade systems represent such products.

5.2 OPEN BUILDING APPROACHES COMPARED

The Supports movement first arose in Europe and Japan in direct response to the effects of post-World War II mass housing. In Japan, issues surrounding the post-war introduction of mass housing were exacerbated by additional factors: medium-rise multi-family housing was a new phenomenon;

the first generation of multi-unit housing stock was quite rigid and of poor quality; and it rapidly grew obsolete.

Throughout the world, post-war mass housing was characterized by the centralization and concentration of formerly dispersed decision-making. This was considered a necessary prerequisite for delivering such large numbers of units quickly. Mass housing also introduced factory-based processes into residential construction. At that time, neither set of processes offered a place for the individual household as a decision-maker. In previous generations, building typologies had helped to preserve a people's way of life, their historical norms and cultural preferences regarding territory, self-expression, entry sequences, and relationship to neighbors. Mass housing, as a rule, recognized none of these. In the name of improving sanitary conditions, rationalizing design and production, it adhered to the impulse of that time to 'reinvent' housing types, ostensibly in relation to machine production.

In western countries, Supports principles were first cast in 1960s political rhetoric of tenant rights and participation, but also in terms of real and effective distribution of control and responsibility. Following the emergence of the term 'Open Building' and of new industrially-produced consumer components for the residential market in the late '70s, a generation of advocates for consumer choice emerged. More recently, additional new and distinct advocacies have become associated with OB: an emerging generation of architects whose cause is sustainable development has shifted the debate toward realizing the potential of OB to dramatically reduce waste. And in Japan, OB is sometimes closely associated with reform of land ownership and tenancy law as well as with the development of better technical systems. Very recently, the Dutch government, following many years of studies, has declared that residential flexibility and choice represent the future of housing. Housing corporations such as Het Oosten are now transferring infill ownership and responsibility to tenants, a move very similar to developments in some housing projects in Japan. Other large Dutch construction companies and development companies now state their intention to develop more consumer-oriented 'one-stop-shopping' capabilities.

5.2.1 Tokyo and Kyoto schools of Supports

Developments toward Open Building follow many diverse and intertwining paths. Although many project solutions actively seek to combine both 'technical' and 'social/organizational' objectives, these two poles or 'camps' do characterize the predilections of various Open Building initiatives and groups.

In Japan, the most active proponents of Open Building have tended to cluster around Kyoto University or The University of Tokyo. It is widely recognized that these two 'schools of thought' are fully complementary, albeit different in orientation. As a result, a number of projects explicitly link the two. The University of Tokyo has tended to address technical issues of Open Building, with signal success. Dozens of developments are springing from The University of Tokyo's lineage of architectural technology innovation and research in collaboration with various government agencies and industry associations. Such developments are best exemplified by the CHS approach. In Next21, for example, innovations include: a sophisticated modular coordination system developed for use by all members of the design team; construction of an innovative skeleton and public infrastructure; installation of new recycling and energy management systems; and development of a new facade kit-of-parts to make subsequent one-at-a-time renovation of units easier and less disruptive than in most multi-story buildings.

The emphasis at Kyoto University has been characterized by its focus on the social restructuring of housing processes to reflect the two basic communities of interest – individual household and the common interest. The latter may be represented by local government housing corporations or by private interests. Process reformation such as Two Step Housing Supply has been incorporated into the Tsukuba Method. The latter, initiated by the Building Research Institute of the Ministry of Construction, has resulted in some of the most important OB projects in Japan. The Two Step Method was also a part of the Next21 project as well as other, more recent developments in the Osaka area. Ongoing work at Kyoto University reveals a second emerging area of emphasis: the search for a uniquely Japanese multi-family housing type. Whereas multi-unit housing to date has been based largely on Western models, this research seeks a direct response to the unique social structure, cultural history, climatic and seismic conditions and other features of Japan.

In Europe, OB-related research activity does not currently approach the level of investment, critical mass or organization present in Japan. Within the Netherlands, OBOM and its followers focus on extending, developing and implementing SAR models of Support/Infill housing, albeit with a more technical orientation. OBOM-related or affiliated work tends to be recognized as the 'official' Open Building. There exist other, overlapping initiatives and movements based on improving and innovating housing, design and construction, but they tend to be perceived as distinct movements. In most European settings, Open Building implementation occurs as a result of entrepreneurial individual action on the part of architects, companies and government research groups, which may then lead to larger initiatives.

Choice for the end user continues to be an explicit goal of almost all realized European OB projects. This tends to be couched in terms of consumer choice rather than political liberation. While group input into the design of common spaces is frequently sought, recent European OB projects have tended to reassert the decision-making role of the designer, who is viewed more as a creator of architectonic form than as a 'skilled enabler.' Facade design has tended to once more become a base building decision. There is, in general, less concentration on political and economic re-structuring than in past decades (with the notable exceptions of Buyrent in the Netherlands and the Tsukuba Method in Japan), and more focus on consumer choice, building and subsystem lifecycles, waste reduction and sustainability issues.

5.2.2 Developments in technical processes and products

Technical developments in OB terms have occurred in two connected spheres of activity, in: 1) 'hardware' development; and 2) changes to con-

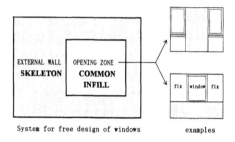

Fig. 5.1 Fixed opening with variable window infill in a Tsukuba Method project. Drawing courtesy of Building Research Institute, Ministry of Construction.

struction processes, permitting and conditions and terms of ownership. OB technical work has been most pronounced in the fields of Support technology (including all building level systems and facades) and infill technology (principally focussed on partitions, mechanical, electrical and plumbing systems particular to the dwelling) and their administrative and physical interfaces.

5.2.2(a) Facade

Technical performance and territorial boundaries between individual occupant and collective are both highly visible and technically demanding in the facade. The facade has consequently become a principle focus of OB research and development. Early explorations and commitment to occupant self-expression in building facades (in La Mémé and Papendrecht, for example), appear to have somewhat waned – with the notable recent exception of Next21 in Japan and some Tsukuba Method projects that explore the use of a fixed opening and variable window inserts.

5.2.2(b) Bathroom and kitchen

Free placement, configuration and selection of parts for bathrooms and kitchens remains a core technical issue of Open Building. From this perspective, the history of Support/Infill in the late 20th century – as exemplified in the preceding milestone case studies – can be viewed as several overlapping generations of infill coming to terms with one central problem: how to untether bathroom and kitchen decisions from the base building. Otherwise stated, efforts to combine individual choice and responsibility, collective coordination and space provision for pipes, ducts and wires and their multiple interfaces in the bath and kitchen represent key technical efforts of OB.

Baths and kitchens have long represented showcases of industrially-produced systems and components. They also represent by far the most technology-intensive habitable zones within the house. Until recently, installing them required multiple site visits by teams of skilled workers in each of several trades to install expert-dependent systems and components. They also account for a major proportion of consumer spending in the residence. Together with restrictions imposed by their mechanical systems, and efforts at further industrialization, these factors have established baths and kitchens as a primary focus of the struggle to establish technical and decision-making rationality, standards and consumer-oriented preferences in housing.

The first generation of Supports in Europe and Japan incorporated 'flexible' column-free spaces, albeit with deep beams in Japan. These designs provided capacity for a variety of normal dwelling functions, spaces and layouts. Within spans of about 4–6m, partition walls could be freely positioned. Bathroom zones, and even individual fixture placements, remained fixed – tightly tethered to the Support's vertical mechanical shafts. Placing kitchen drain lines behind cabinets allowed them to shift somewhat. This approach started in the earliest Open Building work in the 1960s and characterized projects including Überbauung Neuwil Wohlen, La Mémé and 'Davidsboden.' Today, it is still used in projects such as the Free Plan 'fixed' unit in Japan, Pipe-Stairwell Adaptable Housing in Beijing, and VVO/ Laivalähdenkaari 18 in Finland.

In the second generation of projects, the bathroom was raised on a secondary floor. In Japan, beginning with the earliest experimental projects, raised floors were used in the entire dwelling. Within limits imposed by drain pipe diameter and slope, this allowed some slack in the plumbing 'tether' and produced a limited degree of freedom to position rooms and individual fixtures throughout the building, at a cost of additional story height. The use of rear discharge toilets further eased constraints on their location. In Japan, raising the toilet presented no market acceptance problems, because the use of varying floor levels has long been a norm: the entry has traditionally been 10cm below the main floor level of the house. Traditional houses that are entered via spaces with earthen floors placed the tatami level a half meter higher. In both cases, the floor of the traditional ofuro (bath) steps up again.

In Europe, however, raising the bathroom floor was never widely accepted. To avoid this, projects such as Papendrecht and Keyenburg used rear discharge toilets. The large drain line was run in the walls, or along the base of the wall, within a custom-designed cover. The shower and/or bathtub was also raised. Very contemporary and upscale infill kitchen/bath packages with raised bathroom floors continue to feature prominently in current adaptive re-use projects including the Vrij Entrepot warehouse loft residences (1998) located in Rotterdam's Kop den Zuid district. Nonetheless, the use of bath units with raised floors in Europe, such as the Bruynzeel bath unit installed in the Adelaide Road/PSSHAK project (1979), frequently met with strong tenant objections. It has remained subject to criticism for decades.

5.2.2(c) Floor Trenches

Subsequently – and principally in Japan – trenches were formed into the Support floor structure in some projects. This eliminated the raised floor while accommodating the required slope of the drain lines. Both narrow longitudinal trenches and wide trenches were used. They were strategically placed to accommodate a limited variety of bathroom and kitchen locations, both within dwelling units and along party walls. In Japan, the CHS project constructed by Shimizu Corporation in Tokyo exemplifies this approach. The Marelles (1975) project by George Maurios in Paris also exemplifies the use of longitudinal trenches in a grid pattern aligned with structural columns. However, within a process which must battle cost issues presented by the use of relatively expensive value-added infill products, adding complexity to the forming of concrete Supports adds yet more cost.

Beginning in the 1980s and extending into the 1990s, the directional trench began to expand, creating full-width recessed floor zones as part of the Support. This might be accomplished, as in Japan, by inverting the conventional beam/slab structure: upward projecting beams hang the floor slab below, to form floor cavities. Examples include the Hyogo CHS project (1997), several of the Tsukuba Method projects (1998) and the Yoshida Next Generation Project (1998). The Green Village Utsugidai Coop in Tokyo (1993) used a trench extending from one party wall to the other, extending approximately 1/3 of the unit's depth. Within that zone, kitchen and bath units are freely positioned. Next21 uses a trench for distribution of 'public' utilities under the three-dimensional network of public 'streets in the air.'

5.2.2(d) Raised floors

The development of raised floor systems parallels the evolution of Supports. This kind of product was common in many Japanese OB projects as early as the 1970s. In Japan, architects are now offered choices between many raised floor products. Of these systems, there are basically two varieties in wide use. The first is modelled on computer 'access floors': adjustable pedestals with rubber shoes for acoustical isolation support flooring tiles of dense particle board or other rigid materials. Drainage and supply lines and other mechanical systems (cabling and ventilation ducts) are routed underneath this floor. They consequently remain accessible by lifting sections of the floor except where partitions are located. In the second solution, the fit-out includes a 'floor mat.' This is an essentially solid underlayment in which water supply and drainage lines are distributed. Either expanded polystyrene or a secondary layer of concrete or loose granular fill is commonly used. The floor mat is placed as part of the infill. It is thus wholly contained within the unit being served. Generally, it cannot be effortlessly accessed once dwelling installation is complete, particularly when the secondary layer is concrete.

5.2.2.(e) Ceilings

The ceiling as a zone for horizontal distribution of ducts, cabling and other elements of infill has also been a subject for intense study, particularly in Japan and, early on, at OBOM. In the Netherlands, the simplicity of the fit-out and the mechanical systems – in particular, the general absence of air conditioning or ceiling lighting fixtures – made secondary ceilings unnecessary. However, Japan maintains strong cultural traditions of varying ceiling heights relative to room use and proportion, and of placing ceiling light fixtures in each room. These requirements, coupled with heightened requirements for ventilation, humidity control and air conditioning, make secondary infill ceilings almost mandatory in Japan. As a rule, OB projects in Japan utilize a 'dropped' ceiling.

In recent decades, technical research and development in residential infill has focussed on improving systems within the floor assembly, and on implementing flooring/supply distribution subsystems. The goal remains to allow the bathroom and kitchen – and as a result the entire unit layout – to be freely arranged on a unit-by-unit basis independent of dwelling units beside, above or below. In all cases, the incremental steps toward full autonomy of Support and infill – always the goal from the very start – have not been easy.

6

Methods and systems by level

6.1 TISSUE LEVEL

Current Open Building research, publications and realized projects focus primarily on the relationship between base building and infill. As an approach to environmental intervention, OB also fundamentally involves issues of urban design, and although not always explicitly referenced, it builds on a substantial body of related urban scale work. A great deal of applied research on urban tissues, notably during the SAR era, has led to the creation of methodologies for structuring work, establishing the bounds of each profession and recording agreements in combined graphic and text form at the urban scale.

The current text does include one realized urban tissue case study, Beverwaard (1977). Nonetheless, a comprehensive survey of the principles, methodologies, history and state of the art of urban tissue projects based on SAR theory lies beyond the scope of the present work. Some key concepts, issues, working methods and publications associated with urban tissues are discussed in Appendix B.

6.2 SUPPORT LEVEL

Supports provide serviced space for occupancy. As we have seen, Supports can be constructed using many alternative technical systems or materials. In all cases, they provide space to be divided into dwellings, offices, etc.

Supports can be either newly constructed or made from existing buildings. In existing stock, some physical elements are retained and others eliminated to produce an adaptable, 'open' Support building. This process of determination may be accomplished informally. More often it results from capacity analyses in which the design team proceeds iteratively through a series of steps to settle on the best Support design – one with optimum capacity for dwelling or other occupancy variation within the constraints of cost and construction technology.

6.2.1 The systematic design of Supports

Systematic methods of Support design are needed when design becomes too complicated for intuitive common sense approaches, or when the standards and specifications must be made formal and highly explicit. Generally speaking, when a Support has to be designed (either as a new construction or by restructuring an existing building), a systematic approach is needed when: 1) several parties with different interests and skills must jointly make decisions; 2) explicit standards and levels of quality must be agreed upon by a variety of participants; 3) decisions must be made in step-by-step increments, such that each decision leaves open a number of options to be determined subsequently; 4) multiple independent parties must work simultaneously in a coordinated fashion; and/or 5) multiple parties must operate independently to form a coordinated sequence of discrete operations.

Such methods have been developed and are described in *Variations: The Systematic Design of Supports* (Habraken *et al.*, 1976), which states

The basic concept of a support presupposes that at least two participants are making decisions independently and sequentially. First there is the designer of the support who provides an infrastructure in which, at a later date, the resident will create a dwelling using an independent decision making process. What options does the first party leave the second? How can these be analyzed and annotated? Secondly, there is the problem of coordinating the design of the 'infill' which is used to make independent dwellings in the support. These are two separate design processes that operate independently but in parallel, separated physically but not necessarily in time. How can these efforts be coordinated?

The designer of the support operates in a social framework in which his work is tested against generally accepted standards about what consititutes well designed dwellings, as well as the more specific standards of the client, the investor and developer of the building which will be leased or sold to a set of occupants who are not yet known. A least three participants are involved: the designer, the regulatory official, and the client. They have to comply with clearly formulated norms and standards in such a way that these can be effectively applied to compare different series of possible uses of the support. Finally the design of a support involves a number of technical experts: the architect, structural, electrical, sanitary, heating and aircondi-tioning engineers, and builder. As in any other building their various efforts have to be integrated, but in this case they all operate within narrow cost and space limits while having to arrive at a flexible solution. If a predeter-mined floor plan is not available to coordinate their services, other means of communication and coordination are needed.

The basic building systems which form the Support may be grouped in different ways in different national settings. Additional systems may also be appropriated to the Support. Ultimately, all Supports include a structural framework, roof, facade and mechanical systems, as discussed below.

6.2.2 Support Technology

6.2.2(a) Structural frameworks

The cases presented in Part Two virtually all used variations of reinforced concrete structural frames. They were of two basic assembly types: either concrete slab, beam and column; or else concrete slab supported by concrete shear walls. Many use almost entirely cast-in-place concrete of varying

Fig. 6.3 Tunnel-formed Support structure. Photograph by Stephen Kendall.

dimension and morphology. Others utilize composite systems. These in turn include both pre-cast – or post-tensioned – elements and cast-in-place concrete; or masonry bearing walls and concrete slabs – either cast-in-place or utilizing precast planks.

The 'tunnel form' Support is used widely throughout the Netherlands in OB projects as well as in conventional residential construction. It is economical and suited to rapid and systematic construction. It leaves ceilings smooth: in some cases, only a thick paint is needed to finish them. Allowable spans are good for residential spaces and offer a high degree of unit layout variation as evidenced in Papendrecht and Keyenburg.

In Japan, principally because of the severe constraints imposed by seismic design, the skeleton has been much more the focus of attention in the evolution of Open Building. At least six criteria are commonly used in evaluating Support structural systems (Fukao, 1998):

- safety against disaster
- durability
- basic performance as living space
- capacity for enlargement of dwelling space
- flexibility for changing dwelling layouts and interior finish
- adaptability for elderly occupants

Engineering design of S/I housing in Japan has generated many important Support skeleton variants. Among those constructed of reinforced concrete, or of steel beams and columns encased in concrete, are the following types:

1. rigid beam/slab/column type
   a. continuous flat slab over beams
   b. depressed slab between alternating beams (HUDc)

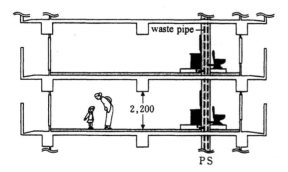

Fig. 6.4 Conventional structural system for residential buildings in Japan. Drawing courtesy of Building Research Institute, Ministry of Construction.

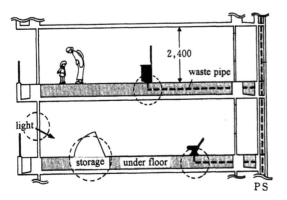

Fig. 6.5 Inverted Slab/Beam Support structure. Drawing courtesy of Building Research Institute, Ministry of Construction.

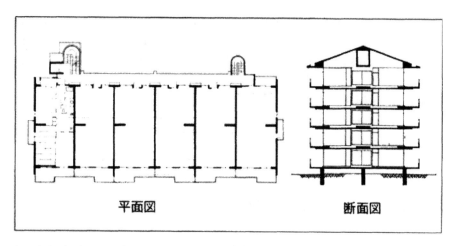

平面図   断面図

Fig. 6.6 Flat beam skeleton. Drawing courtesy of HUDc.

2. inverted slab/beam/column
3. thickened, voided flat slab and columns (Shimizu)
4. bearing wall structure
    a. continuous flat slab over beams
    b. depressed slab between alternating beams (HUDc)
5. flat beam structure (HUDc)
6. z-beam/slab/column type (HUDc)

There are also proposals for all-steel skeletons, for structures that combine steel beams and columns embedded in concrete, and several other kinds. Many experimental systems have been built, tested, and then developed and put into practice.

6.2.2(b) Roof

The roof of a Support reflects cultural and style conventions as well as engineering and weatherproofing concerns. As part of the building envelope, the roof of a multi-unit building is typically constructed as part of the Support. Technical requirements normally prevent any part of the roof from becoming part of infill. In ground-accessed terrace or row projects, however, the roof may be subject to more variation from one house to the next. In the Netherlands in particular, the use of dormers and roof windows is normal, and these elements can be anticipated in the roofs of low-rise Supports, where they are added or modified to suit homeowners.

6.2.2.(c) Facade

Multi-unit residential building facades are generally treated as Support level elements in western nations: as common property, they are assumed to form part of a larger collective decision. Tenants may customarily be entitled to install unique awnings, decorations or even different windows in nations such as China; however, condominium apartments in the West tend to be uniform and may have covenants assuring uniform window treatments. Tenant attempts to control 'their' part of the facade are thus viewed as an aberration in many nations, and may even be associated with buildings which are in social and/or economic decline.

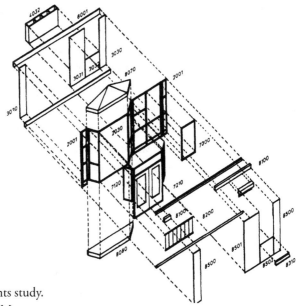

Fig. 6.7 Facade components study.
Drawing courtesy of OBOM.

In redistributing parts of the base building downward to the infill level, division of the facade has proven most controversial. In addressing the public realm, the dwelling facade reflects cultural conventions (having to do with displays of territory, identity and control) and technical requirements (having to do with maintaining structural and enclosure integrity). Questions of who controls the windows, who is legally responsible for them and who pays for repair and maintenance bridge technical and social issues. Are windows to be selected by occupants (as in the Keyenburg, Tsukuba and Next21 projects)? Or are windows common property whose selection and control are issues of the base building, as is conventional in most countries? Work continues at OBOM on The Building Node and the Building Facade Project. These projects aim at developing rules to regulate the connection of building parts as a basis for systematic product development in the industry, but also as a basis for reducing the technical impediments to splitting the facade between Support and infill. (Cuperus,1998)

6.2.2.(d) Mechanical systems

In OB projects, building mechanical systems – relatively recent entrants in the building process compared to structure, roof and facade – are organized on two levels: Support and infill. A significant portion of the mechanical systems – particularly the horizontal distribution of water, drainage, gas, electricity, data and signal wires, heating and cooling – now occur at the infill level. The line capacity of each system, and the technical interfaces among and between them (within and between levels) are critical. They constitute part of the technical maze within 'networked' buildings, as discussed above. In principle, these systems are almost entirely vertical in distribution and orientation at the Support level. Support cabling, ducts, and pipes are normally placed in vertical mechanical shafts inside the building where they run vertically through dwellings or lie in walls between units. They also can be distributed in vertical chases just outside the front door (as in the Hyogo CHS, Pipe-Stairwell Adaptable Housing and other projects), exposed on the outside of buildings, or a combination of both.

6.3 INFILL LEVEL

The story of trends and developments at the Support level is largely about freeing the building's architecture from problems associated with pipes, wires and ducts. At the infill level, it is about the gradual migration of many of the technical and organizational tasks associated with these systems downward to that level. This in turn has led to the rapid growth of infill systems, both partial and comprehensive. Comprehensive residential infill systems are to housing what office fit-out is to tenant space in an office building: they provide everything needed by the tenant to occupy an unfinished space within a serviced shell. They focus exclusively on each individual dwelling as a discrete project, and deliver consumer-oriented products from a single source.

Efforts to rationalize infill have focussed on the spatial distribution and installation of piping, wiring and ducting. Advanced infill systems also rationalize and standardize interfaces for interior partitions, cabinets, plumbing fixtures and appliances. These are the entangled parts which will otherwise cause a 'spaghetti effect' (Van Randen, 1976) in which unpredictable dependencies occur among the many parties involved. These frequently lead to coordination breakdowns and quality control lapses.

The installation of pipes, wires and ducts is of singular importance because these systems must end or begin in coordinated fashion where appliances operate. Modern kitchens and bath/laundry spaces frequently require precise and coordinated pre-positioning of gas nipples, electrical outlets and fixture boxes, hot and cold water rough-ins, water drains and stacks, phone, data and security jacks, television cables, exhaust ventilation and heat and air-conditioning supply and return. In conventional construction, this has required that the position of sockets, vents, jacks, outlets, etc. be fixed during design/development. During early construction, service and supply conduits, cabling and piping are buried within floors and walls, frequently sealed within concrete. As a result, dwelling functions and associated appliances may be locked into place for the life of the building. Any simple change in the location of appliances may require sequenced and coordinated visits by multiple parties, including various trades and inspectors.

Infill systems move these decisions forward in time, to just prior to occupancy. They become variable and easily revised tenant decisions. This approach values consumer choice, and enables subsequent upgrade or inclusion of new technologies, in a world where almost every household installation connects in one way or another to the end of a pipe, wire or duct.

In for-sale development projects, infill systems allow for more efficient customization. They enable 100% market matching in for-rent properties. Infill systems also make one-unit-at-a-time renovation in existing projects easier and make possible rapid, customized residential conversion of obsolete office and warehouse buildings. In allowing for the upgrade of individual building parts without disturbing related systems, infill systems reduce waste and permit the substitution of more advanced, ecologically-sound products and materials as they come to market, without requiring demolition.

In the Netherlands, fundamental research into the technical issues of infill was conducted at the OBOM research group at Delft in the 1980s, resulting in a number of important studies including *Leidingsystematiek* (Vreedenburgh, Mooij and Van Randen, 1990). Among the innovations resulting from this research was the development of the Matura Infill System described below, specifically including the use of zero-slope gray water drain lines. This breakthrough in plumbing technology has very broad implications for the entire building industry. Matura's zero-slope system has been code certified for residential infill applications in both Germany and the Netherlands and has been adopted in the ERA infill product as well. As of this writing, it continues to be examined for possible adoption in Japan.

6.3.1 Comprehensive infill systems

Developments at the infill level – particularly in Europe and Japan – point to a full reengineering of the housing process. The comprehensive infill systems emerging in Europe and Japan share four essential characteristics. They

1) rationalize the production of dwellings; 2) single-source the entire fit-out; 3) certify the infill package as a slab-to-slab prefabricated technical product; and 4) implement advanced information systems to manage the process.

Once such systems have substantially penetrated the market, Supports design will be simplified and relieved of considerable engineering constraints. As a result, it will become far easier for architects designing Supports to refocus on traditional aspects of architectural form and public space, on the building's tectonic qualities, spatial experience, facade and definition of public space and urban character.

6.3.2 Facade Infill Systems

As previously indicated, infill systems are not restricted by technical limitations to the interior of units. Nor is there inherently any issue related to using site-assembled vs. industrially-produced components, although there is good reason to advance the use of industrially-produced high performance facade components. In early Dutch Open Building projects (Papendrecht (1977), Lunetten (1982) and Keyenburg (1984), for example), part of the building facade was determined together with the interior. In Papendrecht, wooden facade frames, adapted from a building method used for centuries in Dutch canal houses, were filled in with windows or solid panels according to the layout of the interior of the dwellings. In Keyenburg, window frame colors were chosen by dwelling occupants from a limited color palette.

Next21 (1994) extended the concept of the facade as part of infill. The street-facing facade is a custom designed system. Design of the individual dwelling includes arrangement of its industrially-produced facade kit of parts, which can also be taken down, modified, and re-utilized in a new configuration. Overall visual coherence and technical performance of the building's facade is made possible by the careful design of the facade system, its components and its rules of coordination, assembly, and subsequent rearrangement.

6.3.3 Infill subsystems

Many of the subsystems, technologies, interfaces and standards which support and enable the development of comprehensive residential infill systems have initially been created for the commercial market. These products are now actively being incorporated into residential projects. In addition to the residential infill companies surveyed in the following chapter, key developments have been pioneered by leading manufacturers including Wieland (Germany) and Woertz (Switzerland), producers of a wide variety of electrical cabling and connectors; DeltaPlast (the Netherlands) and Hepworth (United Kingdom), producers of advanced drainage piping and fitting products; Geberit (Switzerland), a leading manufacturer of plumbing fixtures and related installation products; Sanyo (Japan), a leading manufacturer of heating and air conditioning equipment; and many more.

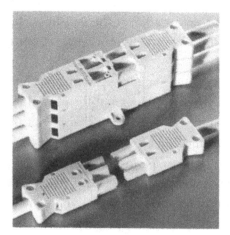

Fig. 6.8 Wieland ST-18 Compact Connector System. Photograph courtesy of Wieland Electric, Inc.

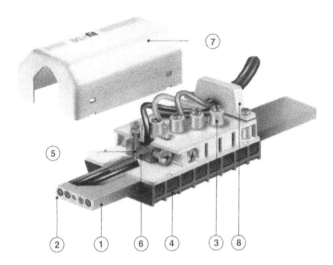

Fig. 6.9 Woertz Flat Cable Installation System. Drawing courtesy of Woertz AG.

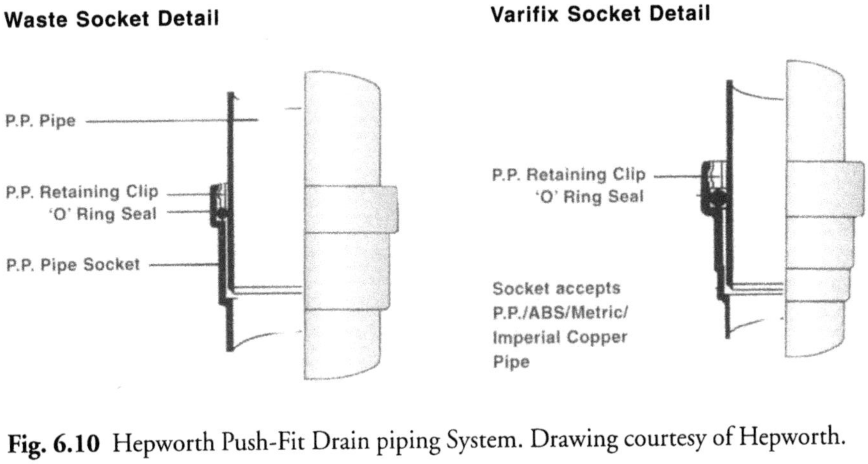

Waste Socket Detail

Varifix Socket Detail

P.P. Pipe

P.P. Retaining Clip
'O' Ring Seal

P.P. Pipe Socket

P.P. Retaining Clip
'O' Ring Seal

Socket accepts
P.P./ABS/Metric/
Imperial Copper
Pipe

Fig. 6.10 Hepworth Push-Fit Drain piping System. Drawing courtesy of Hepworth.

7

A survey of infill systems, products and companies

Matura Infill System
Netherlands and Germany

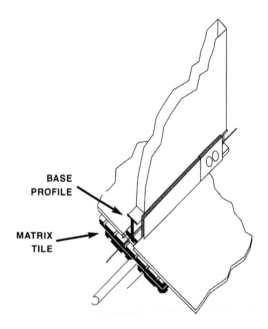

BASE PROFILE

MATRIX TILE

Fig. 7.1 Matura lower system. Drawing courtesy of Infill Systems BV.

Perhaps the most comprehensive infill system to date, the Matura Infill System is a patented product, certified and code approved in the Netherlands and Germany. It was invented in the Netherlands by OB pioneers John Habraken and Age van Randen and brought to market by Infill Systems BV in 1993. Matura, Matura Infill System, Base Profile, Matrix Tile and MaturaCads are proprietary trademarked titles.

The Matura Infill System is a fully prefabricated product. It offers customized just-in-time residential units. The added cost of several of the value-added products used in Matura is off-set by the short completion time for each unit (on average less than 10 working days from delivery of the package to availability for occupancy), quality control, and the ability to offer fully customized units.

Matura is a patented system, registered in the US, Japan, China and the European Community. The patent covers two new products, the Base Profile and the Matrix Tile, and their integration, as well as a software program. The software, called MaturaCads, utilizes state-of-the-art product specification, graphics and accounting principles. This software supports the entire Matura process, from design of the unit to real-time cost estimating, to materials take-offs, to sizing each component for factory production, to labeling and packing in sequence within the dedicated containers assigned to each dwelling unit. Certified multi-disciplinary installers contact building authorities only once installation is completed, since the product certification covers the entire dwelling's infill as a unitary product.

Matura is organized in two subsystems. The 'lower' system uses two patented parts – the Base Profile and the Matrix Tile – to help organize over 23 separate subsystems and thousands of parts already on the market. The Matrix Tile distributes pipes, high voltage cables and ventilation ducts across the dwelling. Pre-cut Base Profiles fit into the top grooves of the Matrix Tile, serving as partition bases and electrical raceways. The Base Profile also allows wiring to be run up into walls and under doorways as required, while also aiding the lower system in accommodating off-the shelf 'upper system' components with standard interfaces. These include partitions, cabinets, appliances, fixtures, door assemblies, etc. Zero-slope gray water drain lines are positioned in the lower grooves of the Matrix Tile while grooves in the upper side contain dedicated 'home run' lines for every fixture. The use of home run lines for water and gas supply is an important principle of many residential infill systems. Home run lines eliminate connections under the floor that require substantial time to install and represent the most frequent source of leaks.

The Matura products and software use a 10/20cm band grid in which each element is assigned a position. Based on the early SAR research, this assures that relations between the thousands of elements can be automatically coordinated. Only at the edges of the dwelling, where the infill meets the Support, is any dwelling-specific measuring, trimming or cutting to fit required. A comprehensive relational data-base contains information on all parts and their possible combinations in sub-assemblies. This data base handles information relative to all parts, making that information easily accessible for fast-track design. In addition, the software includes a graphically-based 'product configurator' integrated with the data-base. Infill design is tested and then recorded with this tool by positioning a symbol of the chosen subsystem assembly within the band grid. This automatically generates all information for costing, ordering, storage, production, container packing, and assembly on site. It does this in predetermined formats suited to each operation.

In projects including Patrimoniums Woningen (1990–), the Matura Infill System is used in both new for-sale and rental units. They are installed one-unit-at-a-time, by a trained three-person multi-skilled installation crew. Installation of an 1100ft² (110m²) custom unit requires eight days on average. To aid occupants with design, a Matura showroom is set up. It includes a dwelling mock-up that illustrates the various technical elements, equipment and range of consumer choices offered. There are also presentations for prospective clients and visitors with catalogues of available fixtures, finishes and cabinets.

Matura's distribution center houses a fabrication shop where infill packages are prepared and the containers for each unit are stocked for delivery to the installation site. Many parts are prefabricated, and some simply pre-cut, to maintain high quality and rapid on-site installation while reducing on-site waste and disorganization. Some parts, such as the Baseboard Profile and the Matrix Tile, are ordered from a supplier to Matura's specifications. Some of these parts are then cut to length and assembled in the distribution center, while other parts are stored for delivery. All parts for a given dwelling unit are loaded into one or two containers in reverse order of their installation sequence on-site. Another container containing all tools required for the job also serves as a construction-site work center for the installation team. These containers remain on site throughout installation of the infill. (Vreedenburgh, 1992)

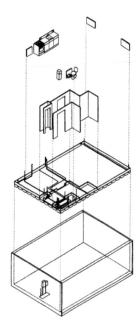

Fig. 7.2 Matrix Tile and 0-slope gray-water drain line installation. Photograph by Stephen Kendall.

Fig. 7.3 Exploded view of Matura. Drawing courtesy of Infill Systems, BV.

ERA and Huis in Eigen Hand (House in Own Hand)
Netherlands

ERA is a large general contractor that entered the market for fitting out dwelling units in existing or new buildings on a 'one-unit-at-a-time' basis in 1998. The ERA Infill System resembles the Matura Infill System, but uses ordinary off-the-shelf products available in the general building products market. It was developed at the request of the Patrimoniums Woningen housing corporation for use in the renovation of a housing project in Voorburg where Matura is also used.

The ERA fit-out product, like the Matura system, is organized in a lower and upper system. To form the lower system, two layers of polystyrene sheets (5cm and 3cm) are laid on the Support floor. Two layers of gypsum anhydrite sheeting are then laid on the polystyrene layers to form a subfloor. Horizontal, zero-slope gray water drainage lines and water supply piping for domestic use and for supplying hot water radiators are positioned within the double layer of polystyrene. Conventional installation technology is utilized to maximize flexibility. Slender acoustically-insulated steel stud and gypsum board 'half-thick walls' are placed over existing walls. As is the case with Matura, the upper system uses off-the-shelf consumer products.

In an elderly care housing project in the Hague, a private housing corporation purchased an outmoded elderly housing project and transformed it into a flexible and consumer-oriented project, using the ERA infill system. An existing 'nursing home' with 360 beds was transformed into an assisted care facility with 160 beds. Twenty existing rooms were transformed into 16 new rooms within a special care facility. Another existing building with 80 rooms for the elderly was gutted and, using infill packages, recombined into 20 apartments. Phased demolition and fit out followed the principles and procedures outlined in *Stripping Without Disruption*, an Open Building study (Dekker, 1997). As a result, the elderly population was able to continuously inhabit the building throughout construction.

The Huis in Eigen Hand product is very similar to the ERA product. It was developed by Karel Rietzschel and first used in 1998 in 50 new dwellings in The Hague for the housing corporation VZOS.

Interlevel
Netherlands

Fig. 7.4 Piping installation under the Interlevel floor. Photograph by Stephen Kendall.

Prowon BV is a development company. It has been developing both office and residential projects based on Open Building principles since 1984. In addition to problems that result from the organizational shortcomings of traditional builders and building processes, there was a perceived lack of inexpensive, easy-to-use and durable products, in particular raised floors and demountable walls.

Prowon developed its own floor and wall system under the trademarked name Interlevel. To market and install these and other products, the Interlevel Trading and Interlevel Building companies were formed. These companies provide the essential organizational tasks of advising owners and users, and producing technical drawings.

Interlevel is an infill product developed initially for the commercial office market. It has also been installed in many residential projects and small businesses. Interlevel's principal component is a proprietary low cost raised access floor. Floor panels are made of high density cementitious wood fiber board. These are supported on wooden frames on adjustable legs at a height of approximately 4–6 inches (10–15cm). The system has approved fire and acoustical ratings under the Dutch building codes. Interlevel also provides the project with conventional polybutylene water piping, electrical conduits and ventilation ducts for placement under this subfloor. Light wood frame or steel stud walls covered by gypsum board are then framed above the subfloor.

Esprit
Netherlands

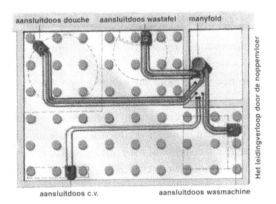

Fig. 7.5 Bathroom floor layer. Drawing courtesy of Esprit.

The Esprit consortium began in 1985 to develop an approach to customized residential fit-out for both new construction and renovation. The system is based on a combination of 'plug-and-play' and Support/Infill concepts (Eger, Van Riggelen, Van Triest, 1991). An early concept sketch shows a homeowner unloading a box from his car, unpacking it and installing a new lavatory in his bathroom with a few simple plug-in steps. Esprit has sought to develop industrially-designed and produced residential infill for the consumer market, including:

- open interior space planning (unrestricted partition wall placement)
- consumer-determined placement of all technical equipment, electrical switches and communication outlets
- open selection of furnishings.

Consortium members include engineering consultants, product manufacturers, and construction companies involved in sanitary and kitchen equipment, heating and ventilation, and partitioning. The Esprit product includes several basic parts:

- a raised floor less than 10cm thick used in bathroom only
- low-slope drain lines and specially-designed drain traps for tub and shower
- bathroom fixtures with proprietary quick connections to drain and water supply lines

- 'plug-and-play' kitchen components in which piping is integrated into the backs of cabinets
- quick-connect water piping for domestic and heating lines
- new space ventilation devices
- a demountable partitioning system
- proprietary power, data and security distribution products (surface and in-partition raceways)
- storage compartments and communication consoles for entry doors.

Part of the Esprit development includes just-in-time distribution throughout the entire design-installation logistics chain. This is linked to ongoing efforts to reduce on-site waste, thus encouraging recycling and more efficient use of materials through a factory production process. A number of demonstration projects have been constructed, including both new buildings and renovations. In 1999, Esprit was reorganized. As of this writing, further demonstration projects have been undertaken.

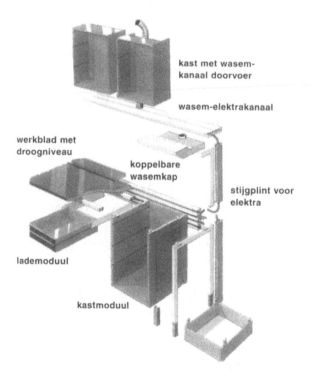

Fig. 7.6 Plug-and-play kitchen. Drawing courtesy of Esprit.

Bruynzeel
Netherlands

Fig. 7.7 Wall system studs being erected. Photograph by Stephen Kendall.

Fig. 7.8 Wall boards installed. Photograph by Stephen Kendall.

Bruynzeel is an established company with a venerable history in the Netherlands. Its reputation was built on trade in forest products hundreds of years ago. Today its products encompass wood products, waterproof plywood, pencils and high quality kitchens for the consumer market.

In the mid-1970s, the company embarked on an ambitious product development program to develop comprehensive residential infill packages (Carp, 1974). Bruynzeel's pioneering system was installed in early OB projects, including Sterrenburg (1977) in the Netherlands and the PSSHAK/Adelaide Road project in London (1979). This system consisted principally of a partition wall system joining milled wooden studs with plastic connector blocks, to which wall panels of particle board were attached. The plastic blocks were used to join vertical studs and horizontal base and top plates. Wiring was surface-mounted in plastic raceways. Drainage piping for the bathroom was installed under a raised floor, and kitchen piping was placed behind the cabinets. Supply water piping was also surface-mounted, using chrome plated piping and carefully detailed brackets. This product was taken off the market in the late '80s because it could not compete with conventional partition products in the market.

Nijhuis
Netherlands

Fig. 7.9 Wall element being installed. Photgraph courtesy of Nijhuis BV.

Nijhuis Bouw BV is a building company founded in 1906. It was Nijhuis who introduced tunnel form building methods in the Netherlands after WWII. Several divisions of the company joined forces in the early 1970s to develop the 4DEE Inbouwsysteem (4DEE Infill System). It included prefabricated interior walls and door frames, facade elements, and roof elements, among other parts developed to quickly finish the company's tunnel-formed concrete shells. 4DEE utilized dimensional coordination principles developed by the SAR, including the 10/20cm band grid.

In 1971, Nijhuis Toelevering BV was formed as an independent company. Between 1971 and 1992, Nijhuis delivered approximately 50 000 dwellings in the Netherlands using 4DEE. The infill system included interior walls, interior door frames and window frames, finishing profiles for baseboards, prefabricated hangers for wash basins and electrical conduits which include boxes. The infill packets are installed at a rate of one day per dwelling, using a two-person crew.

Modular coordination within the system and connections with other systems such as facade, piping and so on rely on the 10/20cm band grid. Wall elements are fabricated in two heights: 2.4m and 2.6m, and in a basic width of 1.2m. This full size panel can be cut down to 30, 60, 90 and 120cm widths.

The 4DEE product is factory-produced for each dwelling unit. Drawings are coded to indicate the placement of each element in the plan. On the basis of these drawings, the factory organizes its production and delivery. On-site, the installation team first fastens a u-shaped profile to the ceiling according to the measured drawings. Tops of walls are inserted into this channel, screw jacks in the panel bottoms are tightened against the floor, and a baseboard installed. Door frames and partitions are installed in sequence. Because the partitions are hollow, electrical wiring can be run in the walls or surface mounted along the baseboard.

Nijhuis Toelevering BV continues to focus on marketing to building contractors. The current product mix includes wooden windows and doorframes; plastic windows and door frames; prefabricated wall-to-wall and floor-to-floor facade elements (including glazed windows and doorframes); and prefabricated roof elements. Products are distributed primarily in the Netherlands for new and renovated dwellings (for a variety of companies including Nijhuis Bouw BV). In the future, Nijhuis expects to incorporate Open Building principles more extensively. A new product development department has been formed under the direction of an expert with broad long-term experience in commercializing residential infill systems. This group is currently testing a concept called TRENTO, which aims to ensure rapid construction while giving buyers many possibilities in choosing unit layouts and facade elements.

Fig. 7.10 Completed installation of 4DEE. Photograph courtesy of Nijhuis BV.

Haseko

Japan

Haseko is a major national construction company in Japan. It is among the largest and most respected housing development and construction companies concentrating on the middle income housing market in Japan. In the early 1990s, Haseko divided its housing production into three main divisions, Support, Enclosure and FORIS (FOR Infill Systems). Haseko's restructuring recognizes profound differences in the types of labor, construction management, and subsystems required by each phase of construction. In their projects, each division is independently responsible for managing costs. Material and labor prices are distributed by division, rather than along conventional technical categories of work.

More than 20 subcontractors are employed in the infill work, including carpenters, finishers, electricians, environmental systems installers, plumbers, kitchen/bath unit installers and so on. Haseko goes directly to parts suppliers to negotiate materials costs for the infill. They then set the material cost to the subcontractors, who control only labor costs. The company can thus monitor labor and material costs precisely. Haseko executes separate contracts with each subcontractor for work on the Support and infill levels, so the price of each can be clearly identified (to the consumer). This is particularly important in conveying to consumers any costs associated with floor plan changes.

Haseko does not use raised or access floors. Supply piping is concealed above a dropped ceiling. Drain lines for the bathtub are easily accomodated under the floor of the unit bath, which is raised, following Japanese convention. Toilets either use a rear discharge tied directly into the vertical stack, or discharge into a waste pipe which then runs along a wall or is concealed in a storage space. Kitchen drain lines extend behind cabinets to a vertical stack. Utility meters are accessible via the public corridor. Domestic hot water is supplied at each unit by a dedicated gas boiler. HVAC is provided by a conventional split system, typical in Japan.

According to Haseko, widespread adoption of Open Building will depend on several factors (Kendall, 1995):

- there must be demand;
- separate infill system providers focussed on consumers must emerge;
- private and public developers must increasingly separate Support and infill contracts; and
- general contractor practices must change.

Panekyo
Japan

Fig. 7.11 Raised floor. Photograph courtesy of Panekyo.

Panekyo (Japan Housing and Components Manufacturers Cooperative) is a national organization founded in 1961. It is headquartered in Tokyo, with seven branches and more than twenty regional offices. Its capabilities include product design and testing, marketing, installation and renovation. For the most part, its products are found in public sector apartment and condominium projects built by national, provincial or municipal housing agencies.

Panekyo has a division called PATIS (Panekyo Total Interior System). It provides complete interior fit-out for multi-family residential units. Patis products include raised floors (7 types), partitions, door units, wall surfaces, ceilings, joinery, built-in components, utilities (including unit bath and kitchen cabinets and equipment), and other interior components. These products have been used in more than 250 000 dwelling units throughout Japan.

Panekyo has sponsored a number of studies of technical concepts specifically oriented to the 'reformation' or rehabilitation of multi-family housing. The reports identify a number of factors bolstering the demand for 'reformed' housing. These include:

- increase of single person households and couples without children;
- increase in the trend to rent rather than buy a condominium unit;
- increased tendency to work at home;
- increase in leisure time and round-the-clock activities;

- demand for housing for diverse life styles and personal expression;
- increased importance of privacy;
- increased demand for enabling DIY activity; and
- increased rehabilitation of residential units.

For builders, factors behind the demand for 'reformed' housing include:

- the aversion of younger workers to dirty, hard and dangerous work;
- increasing worker salaries;
- public dislike of construction and demolition noise and disruption; and
- a shortage of qualified construction workers.

From the standpoint of ecological concerns, reform housing enables environmental protection and preservation, reducing consumption of natural resources and generation of hard-to-dispose-of scrap materials.

Studies by Panekyo suggest a number of strategies for moving infill products into both new Supports and older buildings needing reform. These strategies include developing both multi-skilled installation teams and partial DIY infill. They also include creating a new kind of parts center to sell or lease infill systems; generating a second-hand market for DIY infill components; and developing mechanisms for selling infill components when vacating a dwelling. (Kendall, 1995)

Fig. 7.12 Partition system.
Photograph courtesy of Panekyo.

Sashigamoi

Japan

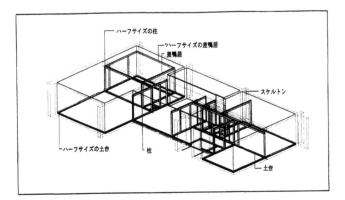

Fig. 7.13 Axonometric view. Drawing courtesy of HUDc.

Sashigamoi specifically refers to the horizontal structural or transom frame member above shoji screen openings in traditional wooden houses. The Sashigamoi establishes a horizontal band throughout the dwelling at a uniform height above the floor. From there, various ceiling heights and decorative screens and walls can be created. This project in Tama New Town near Tokyo was sponsored by HUDc and completed in 1995. Six for-sale units were fitted out. They are 94.55m² in size. The project had four principal goals:

1. to maintain the traditional spatial characteristics of Japanese wooden houses;
2. to maximize the use of wood and natural materials;
3. to use a concrete Support structure with factory-produced interior systems, aided by efficient delivery and installation processes; and
4. to find new markets for Japanese forest products, especially for the less valuable parts of trees.

The six completed units have uniform floor plans. The walls of the unfinished concrete Support are covered by interior 'infill' walls. Spaces are organized so that all drainage and supply piping can be placed under the raised floor of the bath unit. The toilet is positioned immediately adjacent to the vertical piping shaft, with rear discharge.

The meter closet is located at the front door. A 'light court' adjacent to the public stair provides interior rooms with natural light and ventilation, within a deeper-than-usual building.

The same basic concept has been developed to standardize as much of the fit-out as possible in subsequent projects, allowing variations in ceiling height to be accommodated in the zone above the Sashigamoi. Elements from the Sashigamoi to the floor are standard, factory pre-cut and site-installed, while elements above the Sashigamoi are prepared as much as possible off-site, but cut to fit and installed on site. There is, after a fashion, a lower system comprising standard components and an upper system which is somewhat variable for each condition. (Kendall, 1995)

Mansion Industry System (MIS) Infill

Japan

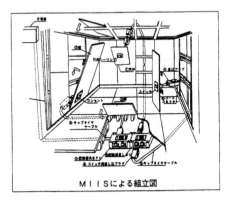

Fig. 7.14 Exploded view of MIS Infill. Drawing courtesy of Daikyo.

The Mansion Industry System was developed as part of the MOC-sponsored Mid- and High-Rise Multi-family Housing Project, the results of which were published in 1992. The project in Fukuoka was developed by Daikyo and the Maeda Development Corporation. Built in 1994, it comprises a 14-story, 250-unit condominium building. Three new systems were used: the Mansion Exterior Industry System (MEIS), the Mansion Interior Industry System (MIIS) and the M(M&E)IS (Mansion Machine and Equipment Industry System).

Dwellings come in six different sizes. Each has a fixed bath unit, toilet and kitchen. The remainder of every dwelling has been custom designed to some extent. All floor plan decisions were made following design of the skeleton. Vertical piping outside the confines of individual units was designed during the skeleton design. However, vertical pipe shaft locations inside the footprint of the units were not set until unit plans for every dwelling were finalized. After all design decisions had been made, construction started. Following one main objective of MIS, material and labor costs were strictly separated.

Exhaust ducts were installed separately prior to interior finish work. Drainage and gas piping were also installed by an independent crew. Several independent teams then worked concurrently, installing each unit in one week. The process began with installation of interior partitions followed by a wiring access floor under which a new 'click-together' cabling system developed by National was used.

Wiring in the project was also distributed in the walls and above a dropped ceiling. Clip-on wall panels were snapped onto the metal stud frames using methods adapted from the automobile industry. Cold and hot water plumbing were installed by the same workers who performed the general interior finishing work, using quick-connect supply and drain lines. All additional parts were organized off-site and brought to the site by the mechanical and electrical contractors.

One of the characteristics of MIS is that parts (piping, cabling, ducts, etc.) are ordered from each manufacturer directly, and delivered just-in-time for installation by the various interior finishing teams. (Kendall, 1995)

KSI Infill
Japan

Fig. 7.15 Raised floor. Photograph courtesy of HUDc.

The KS/I 98 Project was developed by the Housing and Urban Development corporation as one of a long series of experimental housing projects. Its purpose is to demonstrate a new Support construction principle and new infill systems for new and existing housing stock. These systems were developed for use by both HUDc and the private sector. The construction techniques it showcases are intended for buildings up to ten stories. HUDc itself owns 720 000 rental units. Many will require renovation in the coming years.

The present KSI Infill experiment has a number of components newly adopted for this project, as well as a number of previously familiar housing components. The basic subsystems are:

Non-bearing walls
- Outer walls are of dry construction, designed to allow window and exterior door frames to be positioned according to interior layout.
- Party walls that are constructed as Support elements have improved fire and sound isolation performance.

Interior
- A raised floor on adjustable pedestals can be set to a maximum of 30cm above the slab, and is installed after all plumbing is completed.

- Metal stud partition wall frames are pre-assembled and shipped to the site with gypsum board on one side. These partitions are set in place on the raised floor and wired. A gypsum board panel is then fastened to the other side of the partition.
- A horizontal beam is placed at transom height at locations where inner partitions are expected to be installed later.

Mechanical Equipment
- Units have electric heat pumps for heating and cooling.
- The bathroom exhaust uses a constant low-output duct fan which draws exhaust air from the entire dwelling.
- Fresh air intake for the kitchen is placed directly under the kitchen's raised floor.

Plumbing
- The water supply features a header and home run lines for both hot and cold water, running to each fixture and using sheathed resin piping.
- Low-slope gray water drainage lines connect each fixture to a header, which then connects to the vertical drain stack.

Wiring
- Vinyl-sheathed wiring is placed under the raised floor, in partitions and in wiring raceways located at room perimeters in the raised floor.
- Flat wiring, although not yet approved for general use under existing building regulations, is installed on the ceiling.

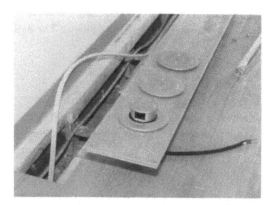

Fig. 7.16 Perimeter electric cabling trench. Photograph courtesy of HUDc.

Finnish infill system developments

Finnish development of residential infill systems is ongoing with increasing funding from industry and government. Different structural solutions have been tried to make altering floor plans and wet spaces and maintaining HVAC systems easier throughout the whole life cycle of the building. Unit bathrooms are now on the market. Light-weight raised floors with easy access for horizontal distribution of service systems are being developed. Experimental flooring systems have employed: expanded polystyrene; light aggregate concrete filling and topping; and boards on steel or wooden 'sleepers.' The access floors are not independently produced components. In most cases, they are assembled by the general contractor using materials from several suppliers.

Modifiable steel framed balcony and external wall component systems exist, but they have not been used as part of the infill. Experiments on facade infill systems have included allowing the consumer a selection of balcony railings of different materials and some options concerning the size of windows.

A demountable partition system with integrated electrical components has been designed for implementation in several projects. However, ordinary lightly framed partitions have sometimes been substituted during construction. The need for increased vertical adjustability in kitchen cabinetry to meet the needs of disabled users, or users with special needs, has accelerated the development of transformable cabinet and table-top systems and their integration with electrical applicances.

To date, complete single-sourced infill systems or a full selection of products that could be used to implement infill packages have not yet been developed in Finland. Experiments with access floors and demountable partitions in prototype systems have not yet achieved fully satisfactory adaptability during use. (Tiuri, 1998)

Chinese infill system developments

Infill system developments in the People's Republic of China are centered first of all on increasing the availability of products (and skilled labor) to a level considered standard in Japan and the West. Such products include metal stud and gypsum board partitions; normal plastic piping for water supply and drain systems; a variety of kitchen and bathroom fixtures, cabinets and appliances; doors, door frames and so on.

The Pipe-Stairwell Adaptable Housing project in Beijing (1994) was built specifically as an experimental Support in which infill systems from various providers could be evaluated. Chinese patent approval was sought for the Matura Infill System and the patents registered in 1998, as efforts continue to find a Chinese company to put it onto the market. The Research Team on Universal Infill Systems for Adaptable Houses has been established in Beijing, joining the Center for Open Building Research at Southeast University in Nanjing in efforts to bring new infill systems and Support designs into the housing market.

Chinese government policy encourages households to purchase dwelling units in multi-family buildings. The general concept of 'open' housing is also encouraged. In open housing in the Chinese context, developers build housing without finishing the interiors. The infill is left for completion by separate contractors in response to individual buyers or as DIY work. As of this writing, organized and systematic methods and products are not yet widely available to make the infill work efficient and cost effective at acceptable levels of quality.

ACKNOWLEDGMENTS

Part Three reflects timely and substantial contributions from many friends and colleagues around the world, both within CIB Task Group 26 and beyond.

Netherlands: In the discussion of infill systems, Age van Randen has been instrumental in providing information about Matura and broad background concerning developments in the European context. John Habraken has also imparted a great deal of knowledge and information, together with electronic graphics. From a unique perspective which combines the roles of OB spokesperson, implementer, client and researcher, Karel Dekker provided information regarding new infill systems and their implementation in practice. Frits Scheublin of HBG informed us about adaptive reuse projects using an OB approach to infill. From OBOM, Joop Kapteijns' work on the Building Node and Ype Cuperus's on capacity studies proved extremely important. Dirk Kuijk at Bruynzeel Keukens provided quite useful background information, as did René van Riggelen and Wim van de Does at Nijhuis. George Kerpel, President of Prowon BV, generously provided information about Interlevel infill products.

Japan: Seiichi Fukao and Shinichi Chikazumi helped obtain and analyze background information on developments on the infill level in Japan, as did Kazuo Kamata (HUDc Research Laboratory). Interviews for a previous study, *Developments toward Open Building in Japan* (Kendall, 1987), included exchanges with Yosuke Yamazaki (Shimizu), Katsuhiko Ohno, Seiji Kurasawa (Haseko) and Shoji Okuda (Ichiura Consultants). They continue to provide crucial information. Hideki Kobayashi at the Building Research Center of the MOC in Tsukuba provided many images and background materials without which the section on the Tsukuba Method of land development could not have been written. Many other experts in Japan have also significantly contributed to developments toward Open Building and to the dissemination of related information; regrettably, only a small portion of their monumental contribution toward worldwide developments and cooperation could be discussed within the constraints of the present book.

Ulpu Tiuri clarified developments at the infill level in Finland, and Zhang Qinnan was instrumental in providing an understanding of the complex situation regarding the development of appropriate infill systems in China.

ADDITIONAL READINGS

Bijdendijk, F. (1999) *Buyrent: the smart housing concept.* Het Oosten, Amsterdam.

Carp, J. and van Rooij, T. (1974) De Ontwikkeling van een taal: het gebruik van een taal. *Plan.* no. 1.

Carp, J. (1979) SAR Tissue Method: An Aid for Producers. *Open House.* **4** no. 2. pp. 2–7.

Carp, J. (1981) Learning from Teaching. *Open House.* **6** no. 4. pp. 2–9.

Cuperus, Y. (1998) Lean Building and the Capacity to Change. *Open House International.* **23** no. 2. pp. 5–13.

De Jong, F.M., van Olphen, H. and Bax, M.F.Th. (1972) Drie Fasen van een Stedebouwkundig Principe. *Plan.* no. 2. pp. 10–52.

Dekker, K. (1997) *Stripping Without Disruption.* KD Consultants, Voorburg, Netherlands.

Eger, A.O., van Riggelen, R. and Van Triest, H. (1991) *Vormgeven aan Flexibele Woonwensen.* Delwel Uitgeverij, 's-Gravenhage, Netherlands.

Fukao, S. (1998) A Study on Building Systems of Support and Infill for Housing in Japan. (ed Wang, Ming-Hung) *Proceedings: Local Transition and Global Cooperation. 1998 International Symposium on Open Building.* Taipei, Taiwan. pp. 100–102.

Habraken, N.J., *et al.* (1976) *Variations: The Systematic Design of Supports.* MIT Press, Cambridge, Mass.

Habraken, N.J. (1998) *The Structure of the Ordinary: Form and Control in the Built Environment.* (ed J. Teicher) MIT Press, Cambridge, Mass.

Kendall, S. (1995) *Developments Toward Open Building in Japan.* Silver Spring, Maryland.

Stichting Architecten Research. (1974) *SAR 73: The Methodical Formulation of Agreements Concerning the Direct Dwelling Environment.* Eindhoven.

Tiuri, U. and Hedman, M. (1998) *Developments Towards Open Building In Finland.* Helsinki University of Technology, Department of Architecture, Helsinki.

van Randen, A. (1976) *De Bouw in de Knoop.* Delft University Press, Delft.

Vreedenburgh, E. (ed) (1992). *De bouw uit de knoop...?/Entangled building...?* Werkgroep OBOM, Delft.

Vreedenburgh E., Mooij M. and van Randen, A. (1990) *Leidingsystematiek.* Werkgroep OBOM, Technical University of Delft, Netherlands.

PART FOUR

ECONOMIC AND ADDITIONAL FACTORS

8

The economics of Open Building

8.1 BASIC ECONOMIC PRINCIPLES

In principle, the economic arguments in favor of Open Building are based on three factors: initial project costs; long-term costs; and total value, including derived social benefits.

It was long assumed that OB could not be justified on a short-term cost basis, or even on the basis of initial construction cost. Support structures build in additional capacity, which results in higher costs. Infill products with value added in the form of embedded knowledge can cost more than conventional interior construction. Accordingly, OB projects were long defended based on enabling tenants significant choice and providing long-term social benefit.

In fact, the case for short-term economic benefits of Open Building is supported by project data. Based on a body of experience derived from well over a hundred realized projects, residential Open Building's first cost benefits have proven to be substantial. Initial savings principally accrue as a result of three factors: 1) the ability to defer and optimize infill investment; 2) reduction of the financing cost for Support construction as a result of expedited completion; and 3) reduction of the cost of construction coordination. Such savings in the construction of the Support may in many cases and markets make total initial project investment competitive with conventional construction, even factoring in the use of advanced OB infill systems. As a result of the demonstrated economies, developers are occasionally purchasing infill systems for installation in speculative residential projects where consumer choice is not a goal. This has been occurring in both new construction and in housing rehabilitation or adaptive reuse projects.

An OB project may be depreciated at the same rate as a conventional building, or it may be depreciated at two different rates: one for the Support, another rate for the infill. Its parts can be upgraded or replaced with less conflict and with more controlled quality than those of a conventional building. Its subsystems are also more efficiently taken down and reassembled or else disposed of. A Support/Infill project built to accommodate long-term change suffers fewer expensive or disruptive renovations and a longer durable life. In brief, however the initial *pro formas* of open and conventional buildings may compare, the savings resulting from Open Building continue to accrue for many decades after savings from value engineering of conventional construction cease. OB's benefits and viability as an alternative to conventional construction are therefore also demonstrated via longer term analyses of building performance.

In traditional financial analyses, building costs are sometimes calculated on a single lump sum investment and product basis. Organizing financial analysis in this way thoroughly discounts the economic effects of duration or tempo of change. Furthermore, all buildings exhibit differential, cyclical tempos of change throughout their useful lives. These rates of change are affected by both internal (building specific) and external (market and economic) factors. At relatively predictable points of a building's useful life, it will – as a whole – experience successive transformations, or remain static for relatively long periods.

Analysis has also quantified how specific uses, systems and parts change at more rapid cycles than others. In many cases, the building would benefit from even more rapid cycles of upgrading and replacement of certain subsystems, if this were possible without substantial demolition, disruption and conflict. To cite examples: Kitchens and their appliances, cabinets and finishes transform somewhat more rapidly than bathrooms; in bathrooms, fixtures and finishes are replaced before the layout, and both kitchen and bath are renovated more frequently than any other zone in the dwelling. The use of electronic and computer-related equipment has grown dramatically, requiring building-wide upgrades to meet increased power demand. Power, data and signal lines and outlets are now routinely run to every space in the dwelling, whereas even a half dozen years ago their presence was significantly less frequent.

To make time-sensitive financial calculation possible, an OB project is first divided into **decision bundles** under the control of various parties:

- developers or builders with short-term interests;
- individual occupants (or their surrogates) with short- to long-term interests; and
- building owners, who generally have mid- to long-term interests (especially public owners; and real estate investment trusts (REITS) in the US).

Financial information related to each party is calculated separately. These calculations are based on the physical elements corresponding to each decision cluster – in Open Building terms, the base building and the infill. Creating an infill decision cluster creates opportunities for investment by inhabitants who would otherwise contribute only depreciation. Clearly, substantial infill investment by renters does dramatically increase both base building and total project cost; but it does so without substantially raising the cost to the owner or any base building investor. It will, however increase equity value for the building owner, and yield a strong and partially quantifiable social dividend over time. Thus, while high investment in infill clearly impacts long-term base building valuation, it does not directly affect the building owner's costs.

Economic incentives or disincentives affecting Open Building valuation vary broadly according to national or regional setting and conventions, investment and tax structure, etc. Nonetheless, certain general valuation principles stand out as key factors that contribute toward Open Building implementation:

1. Valuing long-term factors.   Open Building is most supportable as a building strategy when the long-term consequences of current investment decisions are of considerable concern – and when all negative consequences of short-term investments are fully calculated.

2. Separating decision-making on levels.   The financial benefits of Open Building are optimized when investors, owners, and their project team restructure initial investment. This involves designing and constructing base building and fit-out in distinct decision clusters, by level.

3. Coordinating 'turn-key' fit-out packages.   In consumer-oriented economies, the effectiveness of OB dramatically increases when infill is offered as an integrated product. Like automobiles or computers purchased with specific custom options and configurations, infill is most effective when offered as a single turn-key durable consumer good, purchased for quick delivery and on-site installation by a single responsible party.

4. Certifying infill as a product.   Fit-out systems can be certified as products that satisfy prevailing building codes and product standards. This results in increased product efficiency and in greatly expedited administration of building permits, as well as in reduction or elimination of on-site inspections. Such certification is now in effect for the Matura system in Germany and the Netherlands, for example.

Open Building redefines how parties distribute control and costs. As a consequence, it frequently creates possibilities for innovation in project financing and long-term asset management, as well as more accurate ways to evaluate projects. The following sections illustrate two very different approaches to Open Building finance, in the Netherlands and in Japan.

8.2 TSUKUBA METHOD

Construction of a series of Open Building projects based on the Tsukuba Method began in 1995, following several years of research and development. Created in response to a number of problems related to 'right of use' laws and housing finance, the goal of this initiative is to pioneer a new concept of land ownership and household finance in Japan. It does so while using the Two Step Housing Supply approach. At the time of writing, eight projects including more than 85 units had been built. Significantly, while the first projects were initiated by the Ministry of Construction, the most recent have originated in the private sector.

Life stages of status change associated with housing change have been engrained in Japanese society. It has become conventional for young couples to start out in 'standard quality' dwellings in multi-family buildings. They intend to live in such transitional housing for 5-10 years, then move to a single family residence. There they will stay until old age, at which time they will again move to an apartment.

Unfortunately, as a result of dramatic long-term increases in housing costs, a significant percentage of the population can no longer amass sufficient capital to purchase and maintain a single-family home. These insurmountable housing costs result in large part from Japan's unique land tenure laws, but also from the serious shortage of buildable land for housing.

High costs also frequently prevent landowners in Japan from directly developing their land, or building by themselves. Therefore, a landowner will normally deed 'right of use' of land to secondary parties, who may subsequently construct a building. The landowner legally retains ownership. In theory, landowners therefore maintain the right to reclaim all right of use. In reality, however, 'adverse possession' takes priority. Consequently, the landowner can seldom – if ever – either evict parties who have been granted right of use, or otherwise regain possession. As a result, landowner incentives to sell the right of use are few, while the risks are substantial. Therefore, suitable land on which to build is difficult to find and costly.

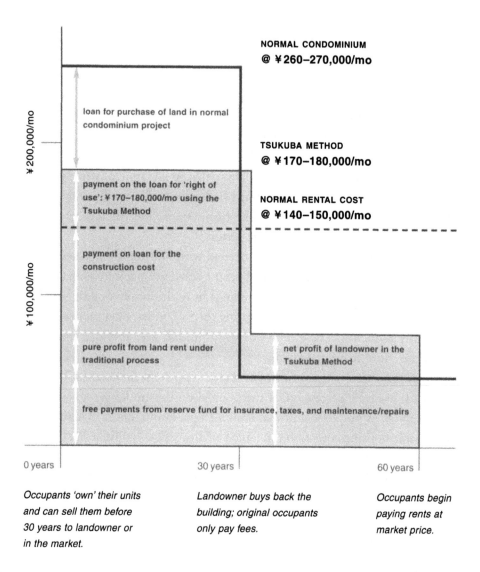

Fig. 8.1 Tsukuba process diagram: Cost Comparison of Three Occupancy Types: Rental, Condominium and the Tsukuba Method. Drawing courtesy of Building Research Institute, Ministry of Construction.

The Tsukuba Method aims to implement a new form of land ownership, and to make it worthwhile and advantageous for households to voluntarily remain long-term in dwellings within multi-family buildings.

In the Tsukuba Method (Fig. 8.1), the landowner in essence leases land to a Cooperative Association, while retaining title to it, somewhat in the manner of a London freehold. In exchange for severely limiting their long-term right of use, members of the cooperative enjoy lowered initial costs and predictable long-term costs. They jointly own the entire project for the first 30 years. In the 31st year, title to the land reverts to the landlord, who, by prior contract, begins leasing it to the households. During the next 30 years, occupants pay only a repair/maintenance fee, as in a condominium, plus a small monthly assessment for leasing the land. At year 60, all units automatically revert to the landowner and are rented at the market rate.

One of the most difficult problems for Two Step Housing had been how to finance the projects. According to Japanese real property law, the ownership of building and land can be divided. The mortgage value of the land is higher than that of building. The Japan Housing Loan Corporation studied this problem, then developed a new leasehold loan which places the mortgage not on the land but only on the building. In creating this new housing loan system to permit implementation of the Tsukuba Method, HLC played an important role as 'co-inventor.' The end result is that it is now possible to borrow 80% of the housing price without a mortgage on the land.

8.3 BUYRENT HOUSING CONCEPT

Buyrent *(Koophuur)* is a newly-implemented system in the Netherlands, in which apartment renters can buy the infill *(Inside)* of their apartments. Infill purchasers enjoy the normal advantages of home ownership, including tax-deductible mortgage interest and the right to modify the interior of their dwelling.

The Buyrent idea was first introduced in 1988 by Frank Bijdendijk, President and Manager of Het Oosten, one of the largest non-profit housing corporations in Amsterdam. Development of this innovative approach, and of the necessary legal and financial instruments to make it work, has taken nine years and three million Dutch guilders (US $1.5 million). Implementation has required careful ongoing coordination with the Dutch Ministry of Housing and Planning (VROM) and the Treasury. Issues such as how to maintain entitlement for rental subsidy for infill purchasers and how to allow occupants to deduct the mortgage interest arising from the purchase of domestic infill had to be resolved. Additional approvals for the experiment were also obtained from the municipality of Amsterdam and the Guarantee Fund for Public Housing (Waarborgfonds Sociale Woningbouw, or WSW). Het Oosten intends to initiate Buyrent not only for itself, but for other housing estate owners as well.

The first Buyrent experiment involved 250 dwellings. It was assessed in 1998. Among the most important findings were the following:

- Buyrent appeals to a broad segment of the market, including all age groups, income levels, kinds of dwellings and kinds of households.
- For lower income groups, Buyrent represents a new option for participating in the housing market.
- Buyrent provides a management instrument for strengthening user involvement in public housing projects.
- The Buyrent instruments, including an innovative way to assess unit costs, have been demonstrated to work in practice.
- The Buyrent product appeals to users for two reasons: it provides ownership; and it offers flexible financing schemes with varied options for setting long-term housing costs.

Ownership offers obvious tax advantages and liberates the user to adapt and improve the interior of the unit. Changes may well affect the assessed resale value of the dwelling infill. But regardless of the financial impact of any adjustments, all dwelling alterations may remain in place when the unit is vacated.

Buyrent offers the owner choices that influence not only total housing costs, but also monthly repayment requirements. Infill purchasers may choose between options which include paying low initial monthly payments with a subsequent gradual increase, or initially high monthly costs that gradually decrease. Based on results from the first Buyrent experiment in Amsterdam, Buyrent will actually lower housing costs for a large percentage of participants. Surprisingly, however, buyer interest is fueled more by the opportunity to tailor monthly costs to individual circumstances than it is by any absolute desire to reduce housing costs.

A National Buyrent Foundation was formed in 1997, as active involvement in Buyrent grew to include the massive Aegon Nederland BV insurance company, and Vesteda Management BV (earlier ABP-Housing-fund. ABP is the largest pension fund in the Netherlands, and one of the largest in the world.)

The evolution of this new approach to housing finance has initially triggered a series of heated and prominently reported debates in legal circles. Among other issues, the Buyrent scheme legally distinguishes ownership of apartment building dwelling from that of collective spaces. Eventually, a majority of lawyers has concluded that Dutch law does permit this, and that this provides a useful distinction. As of this writing, an independent Buyrent finance corporation is being set up. As a turn-key legal, administrative and financial operation, it will allow any tenant in the Netherlands to buy the interior of rented property, borrow money for the purchase and manage the execution of the agreement. Aegon and ABP are partners in this venture. They will also expand Buyrent to serve the luxury market. In high income rental property, the corporation is expected to allow owners to sell their infill directly to new tenants, as condominium owners might.

8.3.1 Legal and financial aspects of Buyrent

Under the Buyrent agreement, ownership of the dwelling unit is separated into the base building and the *inside* (infill which comprises non-load-bearing interior partitioning, floors, technical installations, equipment and finishes). Ownership of the inside is by subscription to an organization of Buyrenters. Buyrenters lease base building space from the building owner. They also buy an infill package and are subsequently in complete control of the dwelling and responsible for its maintenance and for investing in any necessary replacements.

The Buyrenter enjoys an unusual form of purchase protection: The terms of the Buyrent contract cover both the rental of base building space and also the eventual resale of the infill. Upon contract expiration, Buyrenters are obliged to submit their subscriptions to the building owner who, in turn, is obliged to buy them back. The infill's resale value is determined by an independent assessor, following an assessment method developed specifically for Buyrent. The value of the inside depends primarily on the maintenance and quality level of its equipment and utilities, and on the technical quality of the inside components. Additional parameters such as the particular rental policy in effect, and current interest and inflation rates will also influence the sales price. Following the buy-back of a dwelling, the building owner may either rent out the unit conventionally or make it once more available for Buyrent.

Included among the financing mechanisms offered under Buyrent are deferred loans. Based on the 'Private Homes Ruling,' a finding of the Dutch Ministry of Finances, interest paid for such loans is fully tax-deductible, as is the case in traditional mortgages. As a result, the Buyrenter who borrows money to purchase a dwelling can enjoy the same tax benefits as any other home buyer with a mortgage. In future, conventional mortgages may be extended to cover Buyrent.

8.3.2 What Buyrent means to dwelling and building owners

The buy-back assessment method is designed to recognize and reward inhabitant maintenance and improvement initiatives. If the Buyrenter has maintained the inside well and has invested in all necessary routine replacements and upgrades, the property value of the dwelling will rise. Investments to improve the basic materials, finishes and fixtures will be reflected in an increased assessment as well.

Compared to traditional renters, Buyrenters enjoy far greater freedom to customize their dwellings. This holds true for all levels of customization – from hanging pictures to building shelves to rearranging partition walls to renovating baths and kitchens. Buyrenters who invest in customized infill create living space that responds to their needs. Buyrent offers inhabitants whose current dwelling plans no longer suffice a viable alternative to moving. They should therefore have less inclination to look for alternative living locations.

This increases stability throughout the owner's building, and eventually throughout a given neighborhood. Because Buyrenters take care of the infill, the building area most subject to wear and tear, involvement of the building owner in maintenance will be reduced. Risks occasioned by either early and entangled obsolescence or by the introduction of inappropriate uses are both reduced.

Buyrent does not influence the maintenance of the base building. The building owner remains responsible for taking care of this maintenance as before. However, given the dramatically increased involvement of Buyrenters in their own infill, it is assumed that their awareness of and responsiveness to building maintenance issues will increase. They will become more vigilant, more invested and more demanding relative to conditions in their base building.

Initial transaction and administrative costs to the owner are higher in the case of Buyrent than for conventional rentals. These direct costs are expected to be more than compensated for by lower tenant turnover, reduced administrative costs during the contract period, and increased valuation of the building. (Bijdendijk, 1999)

9

Additional trends toward Open Building

9.1 TRENDS IN THE ORGANIZATIONAL SPHERE

Where residential Open Building has successfully taken root and begun to evolve within diverse building cultures and national settings, it has been accompanied by the convergence of certain pressures. It is exceedingly difficult to generalize with regard to cause and effect for all places in the world where Open Building is taking root. Nonetheless, some common threads are observable.

9.1.1 Demand for reduction in friction and disputes

First among these common threads is substantial demand for reduction in friction and disputes among the parties involved in housing processes. In consumer-oriented societies, this includes conflicts between the lifestyle preferences of individual households and the constraints imposed by public regulatory agencies, development organizations or society at various scales. The specific building trades, professionals and technical systems that call for reduction of friction may vary considerably from place to place. Nonetheless, coming to grips with conflicts and disputes represents a universal concern in nations that have invested significantly in Open Building.

With respect to mediating project disruption, dispute resolution offers greater benefits than strictly technical solutions. This ranking takes into account the investment of time and money required to solve conflicts throughout the process of designing, building, managing and rehabilitating residential properties. In addition to conflicts arising from competing or conflicting desires and overlapping turf, disputes arise out of diverse trade and procedural approaches, exacerbated by highly entangled buildings and building processes. Because there exist close interdependencies between all of these categories, a problem (or solution) in one domain greatly affects others.

Many construction and regulatory processes, technical products and skills are incorporated into conventional practice over time. They may remain in use precisely because they help reduce friction between the parties involved. That is to say, building processes, products and skills seem to settle into sustained use because they produce acceptable results while offering stakeholders in the building enterprise maximum independence, convenience, efficiency or freedom.

A good case in point is the detached house on its own plot, serviced by public infrastructure and public streets. In terms of the particular affordances of a suburban lifestyle, no better solution to house building has been found. The detached house easily enables customization. It permits the dwelling to reflect individual preferences regarding style, spatial character, finishes, storage systems, appliances and fixtures. The freestanding house easily accommodates modifications and extensions. Despite the extraordinary degree of systems and procedural entanglement in the conventional single house, the residential typology and its social setting mitigate the inefficiency of technical design: but only up to a point. In many settings, it is implicitly accepted that routine residential renovations drag on for months. During this time, subcontractors representing various trades will parade on and off-site in uncoordinated fashion, altering previously installed work. This in turn causes chains of quality control problems with attendant disputes, rework and cost overruns. But, in the detached house model, there is generally no reverberation into adjacent dwellings.

Given higher density, attached dwellings and pressures for more sustainable urban environment, accommodating the level of inefficiency normal for the detached house becomes more difficult. Where people who live in close proximity nevertheless demand some degree of autonomy and freedom, buildings must accommodate both individual and group demands. With the spread of consumer culture, pressure is mounting for building regulations to stop imposing group choices on the individual. Nonetheless, individual freedom in the manipulation of building elements must not adversely impact community interests. To be freely manipulated at the initiative of the occupant, mechanical, electrical and plumbing systems within the dwelling must be as uncomplicated and as direct as plugging in a computer and as safe to operate as any appliance.

Open Building experts are finding that the gradual accumulation of technical systems in residential construction – a process that has been taking place for over a century and accelerating over the past 40 years – is reaching a crisis point of extreme entanglement and conflict. They see the distinction between base building and infill as a way to restructure the building process, helping therefore to resolve these pressures and conflicts and provide better quality and choice.

9.1.2 Changes in construction, utilities and manufacturing

The construction industry is regulated in part to protect community and individual property investment, but above all to safeguard the public's health, safety and welfare. Consequently, all issues dealing with fire and life safety codes must be resolved satisfactorily. Electrical, plumbing and heating, air conditioning and ventilation systems are regulated for similarly compelling reasons of safety. Progress toward Open Building in the construction industry, while dependent on the particular culture and industry status of each nation, depends upon several key external factors involving regulation, resource systems and manufacturing, outlined below.

9.1.3 Building permit processes and infill systems

Some building regulations and building permit processes must be amended for Open Building implementation to occur, particularly where comprehensive infill systems are to be fully implemented. To begin, applying for building construction permits for a residential Support should ideally follow conventional North American commercial base building practice. In this model, for purposes of obtaining a building permit, capacity for safe and legal subdivision of building floor space according to optional unit sizes can be demonstrated rather than specified. In permitting residential OB projects, for each dwelling size, a number of safe and code-conforming infill variants can similarly be demonstrated for 'blanket' approval. This assures the public of the long-term value and safety of the base building. Its impact in the neighborhood in terms of density, parking, etc. can be also established, nonetheless leaving the occupant or developer free to select a suitable infill.

In the resulting process, permitting is divided into three phases: for the Support, for the building subdivision (parcellation) and for the infill. A phased permit process makes it possible to complete base building permitting without having to establish the final mix of dwelling models, or specific numbers of various sizes of dwellings that will eventually be built-out. This reorganized process also rationalizes public influence on housing on three levels, with decreasing influence as infill decisions are made. It allows building officials to focus public investment of time at the community level of the Support, while freeing them from responsibility in private matters, or intense interference with inhabitants' life-styles, household budgets and so on.

9.1.4 Utility company cooperation

A second basic requirement for successful introduction of infill systems is the cooperation of utility (resource or supply system) companies that provide electricity, telephone, water, drainage and gas and maintain the distribution systems. In an Open Building process, the infill or fit-out part of these systems is the occupant's own responsibility. To maintain a safe building, however, it must be possible to ensure that occupants cannot damage the building's common utility infrastructure via 'their' installation. Decisions must not reverberate disruption to other utility customers. In recent years, this has become possible with telephone connections. Parallel end user independence can also be achieved in the water supply or gas lines by the use of one-way valves. An electronic home safety device is now being developed in the European Community as part of the Home Bus system.

In 1994, the Dutch utilities implemented a new policy to enable Open Building. This permits a multi-skilled **quality certified installer (QCI)** to submit a single certificate of completion for a multiple trade job. The process eliminates the need for separate plan reviews prior to construction, followed by multiple sequential inspections during construction. The utility companies may now independently perform occasional spot inspections at random sites to verify the quality of infill installations.

To be certified as a QCI, a company and its designated workers must demonstrate qualifications to install a certain 'product,' covering an approved scope of work. This certification is granted by an association comprising all of the individual utilities. Such associations in turn developed from pressure from many quarters to reduce conflicts by rationalizing approvals, in particular the control wielded by local building officials. Such control might otherwise obstruct the approval process, increasing costs and causing delays. (Vreedenburgh, 1992)

9.1.5 Manufacturing

Manufacturing has remained vital to construction since the industrial revolution: The history of construction in the 19th and 20th centuries is largely a story of the gradual introduction of manufactured components into ongoing site work. Increasingly, the manufacturing sector has claimed a leadership role in the production of buildings. The 21st century will see a further increase in the use of knowledge-intensive manufactured products. It is critical to determine how these products can support the needs and preferences of both individual households and the communities in which they live.

Open architecture – building with both constancy and change in mind – now requires new initiatives from the manufacturing sector. These must satisfy a new cluster of performance requirements, reorganized by level. Project design teams and clients increasingly recognize that they are not capable of making integrated whole buildings. Nor is it ultimately desirable or productive for them to seek to do so. Therefore, integration by level is key. Manufacturing – and the distribution chains linked to manufacturing – is slowly beginning to recognize and develop for the housing industry a range of products and processes that responds to the same forces earlier witnessed in the office systems industry.

Beyond basic issues having to do with the regulation of manufacturing and industry – safety, recycling, waste reduction, etc. – additional factors now press manufacturing toward product development for the residential consumer market. Among those factors which also favor further development of OB products are the following:

• A growing but underutilized industrial capacity in many nations.
• Effective consumer demand for residential choice at an affordable price.
• A public policy environment – including economic and other structural incentives – in which strong drivers exist for long-term investment in durable built environment which is at the same time capable of change.
• A clear distinction in building regulations and performance measures between matters for which public interests are paramount, and those which can be safely left to the individual. Performance standards for housing must not therefore consider housing in its entirety as a single 'product.' Rather, work should be organized according to levels.
• The development of products, processes, design and construction skills which respond to the above factors and lend themselves to a reduction in technical interfaces and a simplification of information exchange throughout design and building processes.

9.2 SUMMARY

Developments toward Open Building around the world together constitute a wholly unprecedented situation. While the principles of OB are not new and have roots in strong and adaptable vernacular environment, the technical and organizational realities of late 20th century built environment represent a shift of major proportions. This is evidenced by the emergence of new ways of building, new forms of economic analysis required to evaluate time-dependent factors and new and complex forms of distributed ownership, supported by new legal and financial instruments.

We are increasingly dependent on complex and entangled technical systems in everyday life. This strong dependency is coupled with equally powerful expectations that building and dwelling will meet the actual (and changing) preferences and unique needs of each individual, household or community. Such dependencies, when coupled with social expectations, now pose an unparalleled challenge to the practice of substituting expert knowledge for user input, and more generally to the status quo in housing processes. These needs cannot be anticipated, let alone accommodated, by following expert professional judgments about individual preferences.

As the 20th century ends, it is also clear that the production of housing currently suffers from the effects of increasing technical complexity and entanglement in the regulation and control of its processes. Control in the post-mass housing era is distributed, but neither rationally nor evenly. The basic OB principle of using levels to order decision-making and parts, thereby reducing conflict among them and those controlling them matches the increasingly difficult technical and organizational realities of emerging 21st century built environment. Within that context, reducing conflict and distributing responsibility is a fundamental principle on which the growth of hospitable and harmonious buildings and neighborhoods will depend.

ACKNOWLEDGMENTS

Discussion of the economics of Open Building relied to a considerable extent on numerous papers and articles by Herman Templemans Plat and Karel Dekker, as well as on Paul Lukez's (1986) *New Concepts in Housing: Supports in the Netherlands*. John Habraken alerted us to the Buyrent scheme several years ago. He subsequently kept us apprised of developments, and ultimately helped us to obtain materials on the initiative from Frank Bijdendijk, who has shepherded the program from vision to reality. Habraken then further translated key portions of the materials received. Nils Larsson and Karel Dekker helped frame discussion about the link between OB and sustainability issues.

ADDITIONAL READINGS

Bijdendijk, F. (1999) *Buyrent: The smart housing concept.* Het Oosten, Amsterdam.

Dekker, K. (1982) Supports Can Be Less Costly. *Dutch Architect's Yearbook*, the Netherlands.

Hermans, M. (1998) The changing building market as an impulse for flexible, industrialized building. *Proceedings, Open Building Implementation Conference, Helsinki.* June.

Lukez, P. (1986) *New Concepts in Housing: Supports in the Netherlands.* Network USA, Cambridge, Mass.

Tempelmanns Plat, H. (1998) Analysis of the primary process for efficient use of building components. *Proceedings: CIB World Congress on Construction and the Environment.* Gävle, Sweden.

Tempelmanns Plat, H. (1996) Property values and implications of refurbishment costs. *Journal of Financial Management of Property and Construction.* Newcastle, UK. pp. 57–63.

Tempelmanns Plat, H. (1995) Annual cost and property value calculation based on component level. *Journal of Financial Management of Property and Construction.* Newcastle, UK.

Tempelmanns Plat, H. (1990) Towards a flexible stock of buildings: the problem of cost calculation for buildings in the long run. *Proceedings: CIB Congress, Wellington, New Zealand.*

Boas-Vedder, D.E. (1974) *Het Dynamisch Groeiproces: een nieuwe wijze van stadscentrum-ontwikkeling.* Vuga Boekerij, The Hague.

PART FIVE

SUMMARY AND CONCLUSIONS

10

Open Building activity by nation

In recent years, the rate of Open Building implementation throughout the world has continued to increase. Although no single book can do justice to the scope, magnitude and diversity of the efforts currently under way, Part Five identifies ongoing OB activities, developments and dedicated individuals whose work continues to shape Open Building. In what follows, we focus on activities and initiatives that have not been prominently treated in previous sections. We also highlight additional people, research projects, preliminary investigations and publication activities, while briefly listing governmental and other institutions that actively fund or otherwise support developments toward Open Building.

10.1 THE NETHERLANDS

A continuous funding stream in support of innovation and research in housing and building technology has been forthcoming in the Netherlands from a variety of sources, including national ministries and the European Union. OBOM, other research groups, and manufacturers, consortia and individuals working on Open Building implementation have used such funding to support innovative work. OB activities in the Netherlands continue unabated, albeit under a variety of names. In many cases, current advocates of open architecture and OB-like approaches have had only limited association with earlier generations of Supports and Open Building pioneers. Some developments are so recent that little reporting is possible here; some activities are a continuation of developments initiated several decades ago.

10.1.1 Universities

10.1.1(a) Technical University of Eindhoven

Thijs Bax, a founding SAR staff member deeply involved with the development of the tissue model concept, led the Department of Architecture for a number of years concurrent with developing OB-related theory. Paul Rutten conducts Open Building implementation-related research concerning intelligent and easily adapted mechanical equipment in complex buildings around the world. Herman Templemans Plat continues to build on his early ground-breaking work on the economics of Open Building, in which graduate students from various European nations have participated. Jan Thijs Boekholt continues active work in computational methods linked to OB. Jan Westra is active in a variety of initiatives involving innovations in housing, including Open Building, and serves as a board member of the BOOOSTING building innovation group.

10.1.1(b) Technical University of Delft

Ype Cuperus, Director of the OBOM (Open Building Simulation Model) Group continues research and publication activities germane to Open Building methods and implementation. On behalf of OBOM, he provides invaluable liaison with parties around the world seeking information on Open Building and maintains a comprehensive list of Open Building publications, principally in the Netherlands. Also at OBOM, Joop Kapteijns continues to conduct advanced research concerned with systematizing the 'building node' and 'the facade,' taking crucial steps toward making subsystems fully independent. Active in Open Building since the early years of the SAR, and recognized for his urban tissue studies of the 1970's, Kapteijns also realizes residential and commercial OB projects in architectural practice. Age van Randen, OBOM's founder and a former Dean of Architecture at Delft, continues to explore and advance Open Building as a noted expert on infill systems and as a partner in Infill Systems BV. Rob Geraedts teaches and conducts project management studies concerning many topics related to Open Building, both residential and commercial, and was formerly involved in many research projects at the SAR. The Open Building Foundation, founded in 1984, is a network whose mission includes disseminating the ideas, principles and opportunities of OB in practice, and furthering OB through scientific research: its Secretariat and Documentation and Information Center is housed at the Delft University of Technology in the OBOM office. The Foundation recently joined OBOM, Friends of Open Building and other key supporters to hold a symposium and meeting of the CIB Task Group 26 Open Building Implementation in May 1997.

10.1.1(c) Free University of Amsterdam

Koos Bosma (ed), Dorine van Hoogstraten and Martijn Vos are currently completing *Housing For the Millions: John Habraken and the SAR: 1965–2000*, a history of the SAR scheduled for publication in Fall 1999. John Carp, Director of the SAR for a decade commencing in 1976, is providing access to germane information.

10.1.2 Implementers

Fokke de Jong and Hans van Olphen (in the early years working with Thijs Bax as J.O.B. Architects), worked in partnership for several decades, realizing commercial, residential and mixed-use Open Building projects at all levels. De Jong Bokstijn continues designing both newly constructed and renovated adaptive reuse OB projects. Some use conventional infill, others employ state-of-the art comprehensive infill systems such as the Matura Infill System.

Frans van der Werf has continued to teach and consult around the world on Open Building practice at all environmental levels, notably including the US, China and Finland. Van der Werf also continues to build OB projects, in which he links the principles of an open architecture, the pattern language of Christopher Alexander and award-winning ecological design. He is the author of *Open Ontwerpen (Open Building)* (1993).

Henk Reijenga of Reijenga, Postma, Haag, Smit and Scholman (RPHS) Architects continues to work in an OB-oriented approach both in the design of urban tissues and in housing. His projects encompass both new construction and renovation work.

Karel Dekker, Head of the Department of Strategic Studies, Quality Assurance and Building Regulations at TNO Building and Construction Research, is also a coordinator and web master of CIB TG 26. Dekker remains in the forefront of Open Building advocacy in Europe and worldwide, forging links to the sustainability agenda. His work in applied economics has provided the foundation for realizing several recent OB renovation projects.

Eric Vreedenburgh, architect, industrial designer and former OBOM staff member, is constantly exploring in his projects the potential and limits of the OB approach. His work is linked to a specific cultural agenda, which recently resulted in publication in the Netherlands of *The Inevitable Cultural Revolution*.

Frits Scheublin, Director of HBG Engineering, part of HBG Bouw & Vastgoed (Construction and Real Estate), actively supports developments toward Open Building, such as the Vrij Entrepot loft residences in Rotterdam, in one of Europe's premier construction organizations. His efforts are directed toward making construction safer, less polluting, less expensive, faster, less resistant to change yet more durable. Additional architects and builders including: Karina Benraad; Teun Koolhaas; Duinker, van Der Torre, Vroegindewei; De Jager and Lette; Buro voor Architectuur en Ruimtelijke Ordening Martini BV; HTV Advisors BV; Bouwbedrijven Jongen, BV; Architect Office Lindeman; Andre van Bergeijk; and others continue to build residential Open Building projects throughout the Netherlands and beyond. Many others are building similar projects, although not always recognizing that their work fits into the larger story of Open Building.

10.1.3 Research organizations

In addition to groups previously mentioned, a large number of firms and organizations have been active for decades in OB-related research. During the last decade, such activities have involved organizations, projects and products including the following:

- EGM Architecten BV surveyed the effectiveness of Open Building and created policy recommendations for the Dutch national government.
- Nederlandse Herstructurerings Maatschappij (NEHEM) has supported OB implementation.
- DHV Raadgevend Ingenieursbureau BV conducted studies of OB-related infill installations in office buildings, producing landmark cost studies.
- Nederlandse Woningraad (NWR) and Nederlands Christelijk Instituut Volkshuisvesting (NCIV) have both developed policy recommendations related to Open Building.
- Onderzoeksinstituut Technische Bestuurskunde (OTB) has undertaken legal and financial studies of OB.
- The Technical University of Delft continues to conduct research in modular coordination, methodology, system development and building organization.
- The Faculty of Law at the University of Limburg has conducted studies of legal matters related to OB implementation.

A growing number of corporations and individuals are researching, developing and marketing residential infill systems or component subsystems. Among the foremost are: Bruynzeel BV; ERA; the ESPRIT consortium; Infill Systems BV; Nijhuis Bouw BV; Prowon BV; Infra+; and Karel Rietzschel.

10.1.4 Additional agencies and foundations

Various government agencies continue to promote projects and studies to more fully understand OB and related developments. For example, *Flexible Housing*, a recent 32-page publication from the Ministry of Housing and the Environment in the Netherlands, discusses five OB projects and lists product manufacturers, designers and builders associated with them. Early in 1999, IFD-Bouwen (Industrial Flexible Demountable Building), a substantial government subsidy program, was launched. The program incorporates many Open Building principles. It aims to promote the realization of demonstration projects.

The National Buyrent Foundation was created in December 1997. Formed at the initiative of the Het Oosten housing corporation, in partnership with Aegon (an insurance company) and Vesteda Management BV (formerly the ABP-Housing Fund, the largest pension fund in the Netherlands, and one of the largest in the world) to support the marketing, financial and legal implementation of inhabitant ownership of dwelling infill.

In addition, the Netherlands Industrialized Building Foundation, a joint venture between industry, designers and architects, was formed in 1988. The Foundation has published a book called *BOOOSTING in Business: 35 profiles of designers, developers and entrepreneurs in building construction* (the three O's stand for 'Ontwerp, Onderzoek, Ontwikkeling,' the Dutch words for design, research and development.) A number of other private sector organizations and ministries are also stimulating and conducting research and development activities in building innovation.

10.2 JAPAN

10.2.1 Government and other institutions

In Japan, the Housing and Urban Development corporation (HUDc), the Ministry of Construction (MOC), the Ministry of International Trade and

Industry (MITI), the Building Center of Japan (BCJ), the Center for Better Living (BL), the Japan Coop Housing Promotion Association and the Osaka Prefectural Housing Corporation and other local governments continue to play vital roles in experiments on the path toward implementing residential Open Building. Japanese government ministries have traditionally encouraged and coordinated the development of new research, methods, projects and processes through its own agencies, and through many other governmental organizations, third sector (public/private partnership) entities, associations and private corporations. Even in times of market downturn or recession, research involving long-term technology development projects continues to receive substantial support. Such comprehensive projects frequently bring together academic researchers and competing corporations and institutions to form complex and interwoven project groups. Following is a list of major institutional presences in Open Building in Japan.

10.2.1(a) Housing and Urban Development Corporation (HUDc)

HUDc continues the substantial support for Open Building first established in early initiatives such as the KSI project work begun in the 1970s and the KEP project. Kazuo Kamata, Head of the Housing Performance Research Laboratory, and Seiji Kobata, Manager of the Design Division, continue to support and advance Open Building in many ways. The KSI project also serves to showcase infill products and systems from both Japan and the Netherlands.

10.2.1(b) Ministry of Construction (MOC), Building Research Institute

At the Building Research Institute (BRI), Hideki Kobayashi, Director of the Housing Planning Research Unit, has developed and implemented projects laying out new principles of land development based on the Two Step Method. The BRI also funded a landmark study that has resulted in comprehensive documentation of S/I housing implementation throughout Japan, provisionally titled *Skeleton Housing*. At the time of writing, this had not yet become available to the public.

10.2.1(c) Ministry of International Trade and Industry (MITI)

MITI has for many years supported research and development activities related directly to products suitable for export. These products include elements useful in the infill of Supports, mechanical equipment and other interior systems. MITI has, over the years, supported a number of ambitious development projects. Among the most recent is the House Japan project.

10.2.1(d) Building Center of Japan

The BCJ hosted the first exploratory meeting of CIB TG 26 in 1996, under Director Shin Okamoto. BCJ continues to host interdisciplinary meetings and conferences related to Open Building and many other subjects in the building industry.

10.2.1(e) Center for Better Living (BL)

The Center for Better Living is an independent organization established by the Ministry of Construction in 1973. BL continues to evaluate and certify open components for the housing sector.

10.2.1(f) Japan Coop Housing Promotion Association

The Japan Coop Housing Promotion Association exists to promote the idea of cooperative multi-family housing in the Japanese context. Under the leadership of Yoshiyuki Nakabayashi, General Secretary for more than two decades, the Association has promoted many developments closely related to Open Building.

10.2.2 Universities

10.2.2(a) The University of Tokyo

Yositika Utida, during his long and illustrious tenure in the Department of Architecture, was instrumental in many initiatives and projects germane to Open Building. These included broad general initiatives in support of the industrialization of housing production in Japan, the Century Housing

System (1980-) and the Osaka Gas-sponsored experimental project Next21 (1994). He has also instructed and inspired several generations of prominent architects and researchers, first at The University of Tokyo, and later at Meiji University. Tomonari Yashiro and OB fellow traveler Shuichi Matsumura are also at The University of Tokyo, and both very active in numerous Open Building-related research studies.

10.2.2(b) Kyoto University

Kazuo Tatsumi launched the Two Step Housing Supply System and also contributed importantly to CHS and the Osaka Gas Next21 project. His further involvement includes a number of additional initiatives in the Kansai area of Japan, including the development of local industries to fabricate infill elements. Tatsumi's work is furthered at Kyoto University by Mitsuo Takada.

10.2.2(c) Open Building at other universities

Other universities have also fostered the development of Open Building teaching and research in Japan in many ways. They have notably supported the work of a number of leaders in the field, including Seiichi Fukao. Fukao (Tokyo Metropolitan University) is an architect, leading participant in a number of important OB projects and a key figure in CIB TG 26. Projects he has been involved in include Next21 and research investigations concerning many aspects of Open Building implementation over many years. Fukao convenes the Open Building group at the Building Center of Japan.

Additional university-based leaders in Open Building in Japan include the following:

- Masao Ando (Chiba University) has participated in a number of OB-related developments as an architect and researcher over the years. These include feasibility studies of partition systems for residential infill.
- Masaya Fujimoto, Architect (Yamaguchi University) has designed a number of OB and cooperative housing projects.

252 Open Building activity by nation

- Koichi Fujisawa (Shibaura Institute of Technology) has been involved in a number of research and development projects related to Open Building.
- Shin Murakami (Sugiyama Jogakuen University) has studied the renovation of multi-unit housing in diverse national settings.

10.2.3 Additional implementers, architects and researchers

Many additional Japanese architects and researchers have been actively promoting and implementing Open Building for some years, in private architectural practice or government research OB initiatives. They include:

- Shinichi Chikazumi (Shu-Koh-Sha Architecture and Urban Design Studio), was intimately involved in Next21. He has also originated or participated in a number of important OB research and related projects including the Yoshida CHS project.
- Kiyonori Miisho, Architect, has designed many housing projects in conjunction with Professor Utida.
- Seiji Sawada, one of the first proponents of OB in Japan, led the first official Japanese delegation to the SAR several decades ago. He has subsequently led a number of initiatives in research and documentation, has initiated the translation of a number of key European OB texts into Japanese, and serves both as a vital link between Open Building interests in Japan and Germany and as a key member of CIB TG 26.
- Katsuhiko Ohno, with MOC support, has led a number of research groups and projects, including a pioneering project to insert infill units into skeleton housing frames. He also led the recent Mid- and High-Rise Housing research project which involved participation by hundreds of companies and consultants.
- Shigeaki Iwashita (Atias Corporation) has researched Open Building systems throughout the world.

- Makoto Ohnishi (HUDc) is the principal investigator in HUDc's longitudinal tenure study of the Skeleton/Infill Housing System.
- Toshihiko Sugitatsu, Architect, and Naohiro Kawasaki, a researcher (Ichiura Planning and Housing Consultants), have collaborated to develop the Hyogo Century Housing System and the New Housing Supply System for HUDc.

10.2.4 Additional corporate/institutional involvement

Japan's extensive long-term corporate investment in promoting and sustaining research and development efforts directly related to Open Building implementation is clearly unique. Private corporations seeking developments toward Open Building include many of that nation's large and established companies. They include: Daikyo, Haseko, Maeda, Nikkenkei, Osaka Gas, Panekyo, Sekisui, Shimizu, Takenaka, Teisei, Tokyo Gas, Toyota and many others. Public and private corporations have invested in Open Building in partnership with many government initiatives. In so doing, they have been further supported by university-based colleagues too numerous to list. As mentioned above, in many cases, within the intricate and overlapping structure of boards, initiatives and projects, Japan's leading contractors and developers may simultaneously play collaborative and competing roles.

10.2.5 Publications

Japan continues to lead the world in research and resulting publications devoted to Open Building. Many efforts remain somewhat proprietary and inaccessible to the general public. Language and cultural barriers and lack of available funding for translation further hamper efforts to widely disseminate substantial knowledge of the Japanese developments toward Open Building. Among the most prominent and comprehensive recent publications has been the aforementioned comprehensive survey of Japanese Open

Building Projects commissioned by the Building Research Institute of MOC and carried out under the direction of Hideki Kobayashi, with key support from Shinichi Chikazumi and others. A collection organized by Seiji Sawada has also been recently published. This incorporates, among other contributed articles, much of the body of information contained in the recent Dutch publication *New Wave in Building* (Fassbinder and Proveniers, n.d.).

10.3 ADDITIONAL NATIONS

10.3.1 Finland

In recent years, Finland has had to come to terms with its relatively uniform and inflexible housing stock of apartment buildings and row houses. An active search is in progress for housing approaches that are more user-friendly and offer more variety. In that search, architects Esko Kahri, Juha Luoma, Ulpu Tiuri and others have begun a number of Open Building renovation and new construction projects. Working with the Technology Development Centre of Finland (TEKES), Jussi Kautto of the Helsinki City Office Development Unit initiated the recent Open Building Technology Competition in 1999. Housing innovation competitions such as Milieu 2000 and many new initiatives and research and development projects evince significant interest on the part of both public and private sector organizations.

Advanced research and development regarding base building technology, infill systems and financing and construction management in both new housing construction and renovation are ongoing within several organizations. Many organizations and individuals are involved in promoting Open Building ideas. These include Veli-Pekka Saarnivaara and Jukka Pekkanen at the Technology Development Center (TEKES); Veijo Nykänen and Pertti Lahdenperä at the Technical Research Center of Finland (VTT); Ulpu Tiuri and Juhani Kiiras at the Helsinki University of Technology; and Johanna

Hankonen at the Housing Fund of Finland. A consortium of these organizations hosted a meeting of the CIB Task Group 26 in Helsinki in June 1998.

Recent Open Building-related publications in Finland have included Tiuri and Hedman's *Developments Towards Open Building in Finland* (1998) and Lahdenperä's *The Inevitable Change* (1998). Architecture firms directly engaged in Open Building Implementation include: Architect Office Ulpu Tiuri; Esko Kahri and Co.; Architect Office LSV; and Juha Luoma.

10.3.2 United Kingdom

As students at the Architectural Association in 1968, Nabeel Hamdi, current Director of CENDEP at Oxford Brookes University, and Nicholas Wilkinson, editor of *Open House International*, conceived PSSHAK, the first Support/infill project in the United Kingdom. Hamdi's *Housing Without Houses* (1992) subsequently documented much of the debate between 'supporters' and 'providers' in housing. Under Wilkinson, *Open House International*, published at the Development Planning Unit, University College London, has for several decades remained the primary resource for the study and documentation of the theory, methodology and practice of open architecture in its broadest sense, for housing and community development. *OHI's* Urban International Press has recently published a revised English translation of John Habraken's seminal 1961 *Supports.*

In conjunction with Obuild Consultants, Richard Moseley continues to actively promote advanced infill systems in the renovation and new construction markets, and to lead manufacturers on study missions of projects and factories throughout Europe. Moseley is hosting the CIB TG 26 meeting in Brighton in September, 1999 in conjunction with David Gann, Director of the Science and Technology Policy Unit at Sussex University. Gann has organized a number of study missions and research projects related to Open Building, including the conversion of office buildings to housing. His work includes extensive industry- and government-funded research, including a recently authored a book related to OB, *Flexibility and Choice in Housing (1998).*

10.3.3 France

France's Open Building interest and demonstration projects began with Georges Maurios' renowned Support/infill project, Les Marelles (1975). After several decades of hiatus, several new French OB-related projects were completed in the late 1990s by the firm of Georges Maurios. Dutch OB pioneer Frans van der Werf has collaborated with the French Architects Office A.N.C. on the Residence des Chevreuils OB project in France. The Centre Scientifique et Technique du Batiment (CSTB) continues to participate in Open Building through Jean-Luc Salagnac's task group involvement.

10.3.4 Belgium

For many years in Belgium, and now in the former East Germany, The Office of Lucien Kroll continues to produce work that exemplifies how open architecture can liberate mass housing and its occupants from the rigidities of centralized bureaucracy. 'La Mémé,' Kroll's early student housing project at the Catholic University of Louvain, which adopted SAR Supports design methodologies, remains justly renowned, and one of Kroll's most beloved open architecture projects.

10.3.5 Germany

Architects Gutbrod and Rolf Spille designed a number of OB-like projects in the 1970's. The Elementa competition in the 1970's also fostered OB-related developments. Developer George Steinke recently led the successful effort to obtain code-approved certification for the Matura Infill System in

Germany, with the expectation of using it in a large housing project in Berlin. A workshop discussing the application of Open Building methods to the renovation of large concrete panel mass housing took place in Dessau in May 1999. This workshop was organized by EXPO 2000 Sachsen-Anhalt GmbH, under the direction of Gerhart Seltmann and supported by the CIB TG 26 under the leadership of Seiji Sawada, working with Karel Dekker.

10.3.6 Switzerland

From the 1960s on, a continuous and somewhat isolated parallel stream of Open Building and OB-like housing projects has been designed by architects in Zurich, Basel, Lenzburg, and Thun. These include Metron Architects; Bureau ADP; M. Adler, G. Pfiffner, M. Erni; Kuhn, Fischer, Hungerbühler Architekten AG; Büro ADP Architects; Erny, Gramelsbacher and Schneider; Malder und Partners, Architects; Architecture Design Planning; and others. As a diverse group, they have similarly emphasized dwelling unit variability over time, and adjustability to suit user preferences and participation.

Recent Swiss publications in support of Open Building have included *Housing Adaptability Design* (1994), a widely-disseminated ETH Zurich post-doctoral thesis by Jia Beisei. Many OB-like projects have been surveyed by Alexander Henz in his academic publication *Anpassbare Wohnungen* (1995).

10.3.7 People's Republic of China

A number of events in China over the past decade have signaled the national government's increasing interest in reforming methods of housing production. While no official government policies explicitly adopt or promote OB, there is a general awareness that government cannot continue to provide housing in the numbers and at the level of quality now expected. As one result, the climate for innovative housing approaches is improving. Included among the many experiments are a variety of approaches toward realizing a more open architecture, one in which occupants play an increased role.

Open Building has advanced in China as a result of more than a decade of active promotion by a number of individuals. OB and related housing schemes and programs have been developed in Wuxi, Shanghai, Nanjing, Beijing and elsewhere. These have directly resulted from the dedicated work of the following architects (among others):

- Bao Jia-sheng, at the Southeast University Center for Open Building Research and Development (Nanjing) has built several Support/infill projects. Professor Bao continues to organize research, to write and teach in areas related to Open Building.
- Zhang Qinnan, Coordinator, Research Team on Universal Infill System for Adaptable Houses (and former Vice President, China Institute of Architects) has joined with Architect Ma Yun Yu, to realize several OB projects in Beijing and elsewhere. They are also promoting the widespread introduction of infill technology into China through the consulting firm of M & A Architects and Consultants International Co. Ltd.
- Li Daxia has actively advanced OB in Shanghai and continues to do so as part of the M & A consultant group.
- In Hong Kong, Open Building research and publication work is ongoing on the parts of both Jia Beisi (University of Hong Kong) and architect Chen Ke. Beisi's *Housing Adaptability Design* has just recently been published in Chinese (1998).

10.3.8 Taiwan

Ming-Hung Wang (National Cheng-Kung University Department of Architecture) translated into Chinese the basic primer on the design of Support structures *Variations: The Systematic Design of Supports* (N.J. Habraken *et al.*

1978). In collaboration with architect Li-chu Lin (Department of Construction Engineering, National Institute of Technology at Kaohsiung), whose doctoral dissertation researches technical interfaces of Supports, Professor Wang continues to organize symposia and publications dedicated to promoting the implementation of Open Building in Taiwan. Most recently, Professor Wang hosted the Fall 1998 CIB TG 26 meeting and International Symposium on Open Building in key cities including Taipei and Tainan.

10.3.9 North America

Although base building construction and tenant fit-out is now conventional practice in office building and retail mall construction in North America, residential Open Building is not yet commonplace. Nor is the term commonly known. Widespread conversion of obsolete office and warehouse buildings into housing, and parallel conversion of 'live/work' loft projects are precursors to more comprehensive applications of OB principles.

Primarily in the luxury condominium market, new open-building-like projects are being constructed in metropolitan areas, including Dallas, Texas and Boca Raton, Florida. In Florida, Devosta Homes has built tunnel-formed concrete 'four-plexes' (freestanding buildings composed of four dwellings). For rapid installation, Devosta delivers standard interior fit-out packages from an off-site facility, using conventional off-the-shelf infill elements rather than advanced or comprehensive infill systems. In the custom detached house market: Bensonwood Homes President Tedd Benson, renowned writer of *Timber Frame Construction,* the classic textbook on timber frame construction, also explores energy efficient foam-core frame enclosures and has pioneered in the development of open systems for residential timber frame construction.

In Seattle, developer and design consultant Koryn Rolstad joined with the architecture firm of Weinstein Copeland to realize the award-winning Banner Building. That project demonstrated that high profile

Open Building projects in the States can be successful and profitable without relying on new technology.

In Canada, architect Nils Larsson of Natural Resources Canada, Chair of Green Building Challenge 1998, actively promotes the linking of Open Building principles with sustainable development advocacy. Larsson was a key founding supporter of the CIB's TG 26. J.W.R. Langelaan of Langelaan Architects has led in the development of computer software for the ArchiCad program with capabilities directly related to OB design methods.

11

The Future of Open Building

11.1 GLOBAL TRENDS

Profound environmental change, cultural shifts and economic realignments are moving through the societies and building sectors discussed in the preceding chapters. The production of housing continues to be shaped in response, as do the roles played by professionals, public agencies and the public at large.

When the idea of Supports was first conceived four decades ago, the realignments so clearly visible today were only dimly outlined. The Supports idea arose in response to the phenomenon of mass housing, which had come to dominate housing production in the aftermath of the Second Word War. In many nations, centralized control was deemed the most effective way to produce housing; there the existential and social rights of those condemned to dwell in monotonous and monolithic housing estates figured prominently. As Nicholas Wilkinson recounts in the Preface to the revised English-language edition of John Habraken's *Supports* (1999), the international Supports movement took root, not always comfortably, against the backdrop of emerging '60s liberationist movements in the West. Subsequently, explosive urban growth of crisis proportions, and stark monolithic mass housing projects stretching across the horizon became Third World phenomena as well.

In the intervening decades, edge city development and suburban sprawl have defined patterns of environmental growth in many nations. Center cities have sometimes become hollow shells or have burst at the seams with urban in-migration. The prominence of the automobile has grown exponentially everywhere. Followed by the revolution in telecommunications, it has altered the way we live, within communities and within dwellings. Sustainability issues, articulated for decades by a few lonely voices, have now become increasingly important as we learn about the negative impacts associated with conventional development and construction.

For many years, residential construction has stubbornly resisted trends that have transformed other industries, and other sectors of the building industry. However, there is now mounting evidence that evolutions in manufacturing, in technology, in financing, in information management and in the marketplace are fundamentally restructuring the construction, maintenance and renovation of housing. In other parts of the building industry, particularly the office and health care sectors, manufacturing has become a leading force in providing solutions to complex building processes. Although the associated processes of change have been accelerating for some time, professionals in many disciplines are only now awakening to new realities in the housing sector.

Residential construction continues to represent a significant share of nations' overall economies, within which production, investment and profit are all shifting. Along with these shifts, the Supports movement of the '60s and '70s has given way to Open Building. Inhabitants have been recast as purchasers of value-added residential products. As in earlier eras, households enter the marketplace, where they seek to meet their individual needs and preferences, now through the purchase of residential infill products. The introduction of sophisticated manufactured elements is also reshaping the Support level, improving the quality and durability of the shared parts of residential buildings.

Below, we conclude with a summary of significant trends andeffects.

11.1.1 The emerging consumer market for housing infill

The expansion of the market for consumer-oriented residential systems signals the emergence of a distinct infill level. Although sophisticated comprehensive infill products have not yet captured a significant proportion of the construction market for housing interiors, their appearance around the world represents a signal development toward Open Building.

Residential products are now designed, developed and manufactured with increasingly higher added value. Such products offer enhancements in engineered performance, safety, variety and capacity for use and re-use. They are direct-marketed as consumer goods to satisfy an increasingly varied and complex demand, in which personal preference, brand recognition, performance specifications, efficiency, convenience, sustainability, price and monthly installment payments are all major factors.

Building-related industries increasingly seek to stimulate growth of the consumer infill market. In Japan and in many European Community and Scandinavian nations, there is a concerted attempt to establish a market niche for discretionary infill spending on a par with expenditures on entertainment and luxury categories such as travel, electronic equipment and automobile options and accessories. New consumer products and logistical and installation systems entering the market are increasingly compatible with sustainable development principles.

11.1.2 Changing patterns of investment

Investment dollars are shifting from the building site to the manufacturing facility, from new construction to refurbishment, from Support to infill, from material for stock to value-added manufactured components.

The construction industry is adjusting in response to changing patterns of investment: strengthening alliances with real estate development, manufacturing and additional economic sectors.

11.1.3 Advanced information systems

Supported by burgeoning computer networks, by intelligent switching and software and by e-commerce, direct links between customers and producers are multiplying. Via the Internet, consumers now directly purchase customized travel packages, cars and CD's, as well as residential consumer products. Such direct market access is similarly changing relationships between all parties in the construction and development processes, including roles within the design and management professions. In this process, everyone is competing for consumers' discretionary spending.

Intelligent management software has transformed industrial production and other off-site manufacturing. New production methods offer increases in variety, efficiency, quality control, coordination and speed to market, as well as improvements in supply chain logistics, including just-in-time delivery and installation.

The development of comprehensive infill systems software now allows for consumer-responsive design, with real-time calculation of the monthly and long-term financial impact of design choices.

11.1.4 Changes in Supports

The introduction of sophisticated manufactured elements is also reshaping the Support level. For example, facade systems with enhanced energy performance to match the new demand for energy conservation and reduced waste in refurbishment have replaced conventional facades.

The sophistication of mechanical equipment for large multi-tenant buildings – including cabled, piped and ducted systems – is growing. It now provides better environmental control and monitored and self-regulated performance, while also allowing individual tenants to attach and detach equipment without echo effects on other units.

11.1.5 Manufacturing trends

Building product manufacturers are moving to 'time-based' manufacturing with short, 'demand-based' production. Large manufacturers in the US, Europe and Japan have overwhelmingly done so, and other manufacturers are rapidly following.

11.1.6 Trends in construction markets, investment and revenue

Profits are shifting 'upstream' along the value chain. In most industrialized nations, project construction revenues are shifting from the general contractor to the product manufacturer, frequently via the sub-contractor. This allows small project teams, utilizing the inherent capacities of advanced industrial production, to organize projects that are responsive to individual end-user preferences in a 'one-stop-shopping' fashion. Project and investment risk and liability, however, are not necessarily conveyed up the value chain together with revenues.

As links between consumer and manufacturer reshape the construction industry, many intervening steps, services and companies are being eliminated. 'Value chain creep' upward, toward the production of complex manufactured building components, poses a direct threat to some segments of the construction industry. In the US market for office interiors, for instance, Steelcase has now introduced a comprehensive office fit-out system, placing it in direct competition with distributors, interior design firms and subcontractors. Other very large companies are forming consortia to provide more comprehensive open systems in the rapidly changing office sector.

The renovation market is rapidly expanding.   In some nations, investment in the renovation, upgrading and replacement of existing residential stock now equals or surpasses investment in new construction. In many settings, it will exceed new construction for the foreseeable future.

As industry switches its primary focus in response to the market shift, differences between renovation and new construction promote structural change. Renovation is more disruptive for occupants. It is piecemeal in nature and somewhat unpredictable. Renovation also entails technical and organizational complexities not found in new construction.

11.1.7 Changing climate for research

In a climate of market-driven economics, central governments are reexamining the direction and extent of their initiatives in building research and project stimulation.   In the United States, government has long remained absent from direct involvement in restructuring the building industry. Private industry has supported whatever research is done, but at a low level of investment. This is less so in the European Community, where, under the banner of sustainable development, national governments have renewed their interest in a long tradition of building industry research. In Japan, even during economic downturns, the national government continues to stimulate the building and housing industries to find better ways to build for a future environment that is very different from the products of the past half century.

11.2 BUILDING THE FUTURE

During the next 50 years, rates of change in user needs and technical systems in housing are unlikely to diminish. Among the fundamental questions in industrialized nations are: How best to prepare existing residential stock for continuous change? How to increase the efficiency of renovation in the face of increased demands for consumer choice, responsiveness and variety; technical upgrades; a scarcity of labor (skilled and otherwise); and the need for more adaptability in later years of a building's life cycle? Such questions ultimately lie at the core of Open Building discussion. Residential construction must become more efficient, less wasteful and more responsive. It is clear that Open Building is just one development among many in the general movement toward closer alignment – in construction practices – of consumer choice and sustainable development. Information technology, flexible manufacturing, design for re-assembly and other innovations allow high value-added products to be produced in off-site facilities, where they are tailored to end user preferences rather than mass-produced for an anonymous market.

As labor and investment increasingly move away from the site, the methodical use of levels can serve to organize this shift. Concepts and methods associated with levels inherently recognize that a certain portion of the building and its site must be organized to safeguard community interests. Thus, taking the time necessary to build consensus is recognized as an important social decision-making process of the collective site. The private interior spaces of buildings are tied less to community interests and more to the responsibility of the individual user, as are furniture and other personal property or equipment. Interiors can be constructed somewhat independently of external conditions. Infill is highly changeable and also amenable to systematic design, production and installation. The current revolution in building infill, both residential and commercial, is pervasive, but not yet widely recognized.

11.2.1 Environmental diversity

The world-wide market in design, manufacturing and construction is fast consolidating, as are the networks of transportation, information and finance. As design and construction practices and their training and experience grow increasingly international in scope, they export expertise but also proprietary technologies, extending their marketing and support networks. For many, this has led to a fear of either the homogenization of environment throughout the world, or else its MacDonald-ization.

In truth, neither appears to be occurring. Many environments subject to multi-national influence do, in fact, exhibit a high degree of similarity in their exteriors and interiors. They always have. However, this probably expresses a common institutional ethos among decision-makers, more than any subjugation of a powerless market. For design professionals, institutions and clients operating in many territories, the appeal and prestige associated with unifying control and standardizing environment is undeniable.

Yet the world's emerging consumer-driven housing markets remain subject to singularities of site and microclimate, to local codes and building culture, to national and local and class-bound customs and traditions as well as to personal preferences. At the root of commonly-occurring shifts in the building industry is a very local process of sorting out what is generally applicable and what must remain inherently local: The same culling will occur in the context of any regional or national marketplace, each of which presents a microcosm of similar movements occurring internationally. Both in Japan's organized quest for a modern and uniquely Japanese housing typology and in the North American boom in luxury timber frame construction utilizing advanced infill systems, there is a search for balance. On the one hand, there is the desire to participate in the cumulative tradition and knowledge of generations of builders. On the other hand, there is the desire to optimize the benefits of housing, including incorporation of state-of-the-art subsystems in response to changing requirements and preferences.

Given a choice between new products made available on the open market and participation in shared building culture, it becomes clear that inhabitants demand both.

11.3 CONCLUSION

In surveying the state of the art of residential Open Building worldwide, it is clear that housing projects based on Open Building principles are breaking ground in increasing numbers, from Nanjing to Osaka, Seattle to Paris and Amsterdam to Helsinki. The residential construction industry is variably, and not always eagerly, transforming in the process. Where Open Building is not yet economically or organizationally viable, government agencies and private corporations are slowly recognizing that it is a beneficial – and probably inevitable – trend. Some have therefore invested for the long-term, spending many billions of dollars on Open Building research, development and implementation.

Urban-level Open Building strategies, such as the SAR Tissue Methods and principles, have produced useful studies and projects. To date, few of those methodologies have entered widespread practice. As the worldwide effects of uncontrolled urban growth increasingly come to the fore, this may change. The need for effective methods to manage urban design within sometimes highly charged and political processes may focus attention once more on the sorts of methodical rigor, capacity to connect to historical urban fabric, and foundations for team work offered by the SAR Tissue Methods.

The proceeding chapters have also shown that at the base building and fit-out levels and their interface, significant technical, procedural, financial, legal, social and code-related hurdles remain to be overcome. New initiatives are needed to clarify and extend the basic principles of Open Building. Future methods must:

- more effectively organize and coordinate work by different parties on different levels;
- reorder technical interfaces so as to reduce conflict and ease replacement and substitution of parts; and
- lead to the realization of better, more adaptable, durable and sustainable buildings and neighborhoods.

In the meantime, ongoing developments – in residential, mixed-use and commercial Open Building alike – are clearly of increasing interest to parties engaged in design, construction and allied industries throughout the world. In all likelihood, developments toward Open Building will progress in two distinct situations:

Where previously centralized and unified control of housing activity continues to decentralize, players in the building industry may gradually adopt Open Building to remain in control of increasingly complex projects and processes. Such developments will be most apparent where building processes become overly burdened by technical complexity and where there is increasing demand for better long-term asset management in the face of relentless change. To solve the problems accompanying a break-up of centralized patterns of work and control, bringing new conflicts and increased risk, methods such as those embodied in Open Building will be in demand.

Open Building may also flourish in situations where strong expectations for individual consumer choice collide with problems of social inequity and the procedural and physical limits of conventional construction. There as well, attempts by people in all walks of life to realize personal preferences in the dwelling and in the work place may find limits imposed by technical complexity, procedural barriers and demand for long-term value in the face of constant change. Such bottom-up demand for variety and equity and responsibility will also bring methods such as Open Building to the fore.

Appendix A

Realized Open Building and related projects by nation

AUSTRIA

1968   Saalwohnungen, Vienna
Architect: Kratochwil

1972   Dwelling of Tomorrow,
Hollabrunn
Architect: Dirisamer, Kuzmich,
Uhl, Voss and Weber

BELGIUM

1974   'La Mémé' Medical Student
Housing, Catholic University
of Louvain, Brussels
Architect: Office of Lucien
Kroll

CHINA

1956   Housing Project, Tianjing
Architect: Peng, Qu

1987   Support Housing, Wuxi
Architect: Bao

1990   House #23 of the Huawei
Residential Quarter, Beijing
Architect: Beijing Building
Engineering Design Co, Ltd.

1991   Huawei No. 23, Beijing
Architect: Zhou, Zhang and
Zhou

1992   Experimental House No. 13,
Block 15, Kangjian Residential
Quarter, Shanghai
Architect: Liu, Wan, Ye of the
Shanghai Light Industry Design
Institute

1994   Pipe-Stairwell Adaptable
Housing, Cuiwei Residential
Quarter, Beijing
Architect: Ma and Zhang, M &
A Architects and Consultants
International Co.

1994   Flexible Open Housing with
Elastic Core Zones, Friendship
Road, Tianjin
Architect: Huang Jieran +
Tianjin Housing Estate
Development Holding
Corporation

1995   Partial Flexible Housing in
Taiyuan, Shanxi Province
Architect: Ma and Zhang, M &
A Architects and Consultants
International Co.

1995   Beiyuan Residential Quarter in
Zhengzhou, Henan Province
Architect: Ma and Zhang, M &
A Architects and Consultants
International Co.

1998   Partial Flexible Housing in
        Beiyuan Residential Quarter,
        Zhengzhou, Henan Province
        Architect: Ma and Zhang, M &
        A Architects and Consultants
        International Co.
1998   Housing Tower, Pingdingshan,
        Henan
        Architect: Ma, Zhu, Sun of
        Section #7, China Building
        Standardization Research
        Institute

ENGLAND
1975   PSSHAK: Stamford Hill,
        London
        Architect: London GLC
        (Hamdi, Wilkinson)
1979   PSSHAK: Adelaide Road,
        London
        Architect: London GLC
        (Hamdi, Wilkinson)

FINLAND
1995   VVO/Laivalahdenkaari 18,
        Helsinki
        Architect: Oy Kahri Architects
1997   Sato-Asumisoikeus
        Oy/Laivalahdenkaari 9, Helsinki
        Architect: Eriksson Arkketehdit
        Oy (Petri Viita)
1999   Tervasviita Apartment Block,
        Seinäjoki
        Architect: LSV Oy/Juha Luoma

FRANCE
1975   Les Marelles, Paris
        Architect: Maurios

1990   Residence des Chevreuils/Paris
        Architect: Architect Office ANC
        + Van der Werf

GERMANY
1903   Skalitzerstrasse 99, Berlin
        Architect: n.a.
1927   Häuser am Weissenhof,
        Stuttgart
        Architect: Mies Van der Rohe
1970   Haus am Opernplatz, Berlin
        Architect: Gutbrod
1972   Elementa '72, Bonn
        Architect: Offenbach, PAS
        Architects and Town Planners
1973   Project 'Steilshoop,' Hamburg
        Architect: Spille, Bortels
1973   MF-Hause 'Urbanes Wohnen,'
        Hamburg
        Architect: Spille UA
1979   Feilnerpassage Haus 9, Berlin-
        Kreuzberg
        Architect: Randt, Heisz, Liepe,
        Steigelmann

JAPAN
1980   KEP Maenocho Project,
        Itabashi-ku, Tokyo
        Architect: KEP Project Team,
        Housing and Urban
        Development corporation
1982   KEP Estate Tsurumaki, Tama
        New Town, Tokyo
        Architect: KEP Project Team,
        Housing and Urban
        Development corporation
1982   KEP Town Estate Tsurumaki,
        Tama New Town, Tokyo

Architect: KEP Project Team, Housing and Urban Development corporation

1982 Senboku Momoyamadai Project Sakai-shi, Osaka
Architect: Osaka Prefectural Housing Corporation + Tatsumi Laboratory and Seikatsu-kukan Keikaku Jimusho

1983 Estate Tsurumaki andTown Estate Tsurumaki, Tama New Town, Tokyo
Architect: Housing and Urban Development corporation, Kan Sogo Design Office, Soken Architects, Alsed Architects

1983 C - I Heights, Machida, Machida-shi, Tokyo
Architect: Takenaka Corporation

1984 Pastral Haim Eifuku, Suginami-ku, Tokyo
Architect: Shimizu Corporation

1984 Cherry Heights Kengun, Tokyo
Architect: Kumamoto Prefecture Public Housing Corp + Ichiura Architects

1985 PIA Century 21, Kanagawa
Architect: Shokusan Housing Corporation

1985 L-City, New Urayasu, Chiba
Architect: Haseko Corporation

1985 Tsukuba Sakura Complex, Tsukuba
Architect: Alsed Architects and Urban Designers

1986 'Free Plan Rental Project,' Hikarigaoka, Nerima-ku, Tokyo
Architect: Housing and Urban

Development corporation, Kan Sogo Architects

1986 CHS Project: Terada-machi Housing, Osaka
Architect: Osaka City Public Housing Supply + Yasui Architects

1987 MMHK CHS Projects: Chiba
Architects: Ohno Atelier, Kinoshita + Hosuda + Minowa Real Estate + Marumasu Ltd.

1987 Yao Minami Housing Osaka
Architect: Osaka City Public Housing Corp. + Itagaki Architects

1987 Yodogawa Riverside Project #5 Osaka
Architect: Osaka City Public Housing Corp. + Tohata Arch.

1988 Villa Nova Kengun, Kumamoto
Architect: Kumamoto Public Housing + Ichiura Architects

1988 Rune Koiwa Garden House, Tokyo
Architect: Haseko Corp.

1989 Senri Inokodani Housing Estate Two Step Housing Project, Osaka Architect: Osaka Prefecture Housing Agency + Tatsumi and Takada and Ichiura Architects

1989 Saison CHS Hamamatsu Model, Shizuoka
Architect: Ichijo Construction

1989 Centurion 21, Toyama
Architect: Taiyo Home

1993 Green Village Utsugidai coop project, Hachioji

Architect: Housing and Urban Development corporation + Han Architects Office

1993– House Japan Project, Tokyo
Architect: Ministry of International Trade and Industry + Matsumura, Tanabe

1994 Next21, Osaka
Architect: Osaka Gas + Next21 Project Team

1994 MIS Project/Shirakibaru Project, Fukuoka
Architect: Daikuyo + Maeda Development Group

1994 Takenaka Matsuyama Dormitory Project, Osaka
Architect: Takenaka Corporation

1995 Sashigamoi Interior Finishing Method, Tama New Town, Tokyo Architect: Fujimoto

1995–7 Action Program for Reduction of Housing Construction Costs, Hachioji-shi, Tokyo
Architect: Housing and Urban Development corporation Design Section

1996 Block M1-2, Makuhari New Urban Housing District, Chiba
Architect: Shimizu Design Department + RTKL

1996 Tsukuba Method Project #1 Two Step Housing Supply System, Tsukuba-shi, Ibaraki
Architect: Building Research Institute, Ministry of Construction + Takenaka Corporation

1996 Tsukuba Method Project #2, Two Step Housing Supply System. Tsukuba-shi, Ibaraki
Architect: Building Research Institute Ministry of Construction + Ataka Corp.+ Tokyu Koken Corp.

1997 Hyogo Century Housing Project, Hyogo Prefecture
Architect: Hyogo Prefecture Housing Authority + Ichiura Consultants

1997 Elsa Tower Project, Daikyo Corporation, Tokyo
Architect: Takenaka Corporation, Tokyo Design Department

1997 HOYA II Project, Tokyo
Architect: Taisei Prefab Corporation Design Department

1998 Yoshida Next Generation Housing Project, Osaka
Architect: Osaka Prefecture Housing Corporation and Construction Committee of the Next Generation Housing for Municipal Housing Corporation (Tatsumi, Takada, Yoshimura, Chikazumi)

1998 Matsubara Apartment/Tsukuba Method Project #3, Tokyo, Japan
Architect: Building Research Institute, Ministry of Construction + Takaichi Architects + Sato Kogyo Corp.

NETHERLANDS

1935 Complex 'De Eendracht,'
Rotterdam
Architect: Van der Broek

1969 Housing Complex, Horn
Architect: Van Wijk and
Gelderblom

1970 Six Experimental Houses,
Deventer
Architect: Van Tijen, Boom,
Posno, Van Randen

1973 Rental Housing, Genderbeemd
Architect: Van Tijen, Boom,
Posno, Van Randen

1973 MF-Haus, Rotterdam
Architect: Maaskant,
Dommelen, Kroos

1974 Vlaardingen Holy-Noord
Architect: Werkgroep KOKON

1974 Social Housing in Assen-Pittelo
Architect: Van Tijen, Boom,
Posno, Van Randen

1975 Social Housing, Stroinkslanden
(Zuid Enschede)
Architect: Van Tijen, Boom,
Posno, Van Randen

1975 Social Housing, Zwijndrecht
(Walburg II)
Architect: Van Tijen, Boom,
Posno, Van Randen

1975 Housing in Kraaijenstein
Architect: Van Tijen, Boom,
Posno, Van Randen

1975 Zutphen - Zwanevlot
Architect: Van Tijen, Boom,
Posno, Van Randen

1977 Sterrenburg III, Dordrecht
Architects: De Jong, Van Olphen

1977 De Lobben, Houten
Architect: Werkgroep KOKON

1977 Papendrecht, Molenvliet
Architect: Van der Werf,
Werkgroep KOKON

1979 Haeselderveld, Geleen
Architect: Wauben

1980 Housing Project, Beverwaard-
seweg, Ijsselmonde
Architect: Kapteijns + Interlevel

1980 Housing Project, Tristanweg,
Ijsselmonde
Architect: Kapteijns + Interlevel

1980 Tissue/Support Project, Leusden
Center (Hamershof)
Architect: Van der Werf

1982 Lunetten, Utrecht
Architect: Van der Werf,
Werkgroep KOKON

1982 Baanstraat, Schiedam
Architect: Kuipers, Treffers and
Polgar, ARO Consultants

1982 Dronten Zuid
Architect: INBO, Woudenberg

1982 Niewegein
Architect: Bureau Wissink and
Krabbedam

1984 Keyenburg, Rotterdam
Architect: Van der Werf,
Werkgroep KOKON

1987 Tissue Project, Claeverenblad/
Wildenburg
Architect: Van der Werf

1988 Berkenkamp, Enschede
Architect: Van der Werf,
Werkgroep KOKON

1989   Housing Project,
Zestienhovensekade, Rotterdam
Architect: Kapteijns + Interlevel

1990   Support/Infill Project,
Kempense Baan, Eindhoven
Architect: De Jong, Van Olphen

1990–   Patrimoniums Woningen
Renovation Project, Voorburg
Architect: Reijenga, Postma,
Haag, Smit and Scholman
Architects + Matura Inbouw

1990   232 experimental houses,
Zwolle
Architect: Benraad

1991   Flexible Infill Project,
Eindhoven
Architect: De Jong and Van
Olphen + Matura Inbouw

1991   Meerfase-Woningen, Almeer
Architect: Teun Koolhaas
Associates

1991   Schuifdeur-Woning, Amsterdam
Architect: Duinker, Van der
Torre

1992   Patrimoniums Woningen New
Dwellings, Voorburg
Architect: Reijenga, Postma,
Haag, Smit and Scholman
Architects + Matura Inbouw

1994   42 student apartments, former
office building, Rotterdam
Architect: Benraad

1994   Housing Project, De Raden,
Den Haag
Architect: Kapteijns and Bleeker
+ Interlevel

1995   53 'Houses that Grow,' Meppel
Architect: Benraad

1995   Elderly Care Housing, Eijken-
burg, The Hague
Architect: Vroegindewei and
ERA Bouw + ERA Infill

1995   Housing Project, De Bennekel,
Eindhoven
Architect: Kapteijns and Bleeker
+ Interlevel

1996   Gespleten Hendrik Noord,
Amsterdam
Architect: De Jager, Lette Archi-
tecten

1997   28 Open Building houses,
Nieuwerkerk aan de Ijssel
Architect: Benraad + Prowon/
Interlevel

1997   Puntgale Adaptive Reuse
Project, Rotterdam
Architect: De Jong, Bokstijn

1997   6 Support/Infill Houses, Ureterp
Architect: Buro voor
Architectuur and Ruimtelijke
Ordening Martini + Matura
Infill

1998   The Pelgromhof, Zevenaar,
Gelderland
Architect: Van der Werf

1998   Support/Infill Project of 8
Houses, Sleeuwijk
Architect: De Jong, Bokstijn +
Matura Infill

1999   45 Three-room-houses in former office, Delft
Architect: Benraad

1999   VZOS Housing Project, the
Hague
Architect: HTV Advisors BV +
Huis in Eigen Hand Infill
System

SWEDEN

1950 Wohnblock, Göteborg
Architect: William-Olsson

1954 Flexibla Lägenheter, Göteborg
Architect: Tage and William-Olsson

1955 Mäander-Seidlung, Orebro-Baronbackarna
Architects: Ekholm, White, et al.

1959 Kallebäckshuset, Göteborg
Architect: Friberger

1960 Apartment Block in Göteborg
Architect: William-Olsson

1966 Diset Project, Uppsala
Architect: Axel, Grape and Konvaljen

1967 Housing Project, Kalmar
Architect: Magnusson, Marmorn-Porfyren

1967 Orminge, Stockholm
Architect: Curman, Gillberg

1971 Housing Project, Kalmar
Architect: Magnusson, Marmorn-Porfyren

1976 Öxnehaga, Husqvarna
Architect: n.a.

SWITZERLAND

1966 Überbauung Neuwil, Wohlen
Architect: Metron Architect Group

1974 Überbauung Döbeligut, Oftringen
Architect: Metron Architect Group

1986 Schauberg Huenenberg, Hünenberg
Architect: Büro Z-Architects

1990 Hellmutstrasse, Zürich
Architect: Architecture Design Planning

1990 Herti V, Zug
Architect: Kuhn, Fischer, Hungerbühlere Architekten AG

1991 Hellmutstrasse, Zurich
Architect: Büro ADP Architects

1991 Davidsboden, Basel
Architect: Erny, Gramelsbacher and Schneider

1993 Luzernerring, Basel
Architect: Malder und Partners, Architects

1994 Überbauung 'Im Sydefädeli,' Zürich
Architect: Architecture Design Planning

1994 Wohnüberbauung Wehntaler-strasse-in-Böden, Zürich
Architect: Architecture Design Planning

1995 Muracker, Lensburg
Architect: Pfiffner, Kuhn

UNITED STATES

1994 Banner Building, Seattle
Architect: Weinstein Copeland Architects

Appendix B

The SAR Tissue Method

Urban tissue represents the scale of recognizable and commonly understood neighborhood character, combining discernible patterns in the ordering of public space, buildings and activities. It defines interrelations at a scale smaller than the urban structure but larger than the single building, a scale where a large number of discrete architectural interventions are integrated with streets and public spaces to fill voids in urban structure. Within such tissue, variations serve to reinforce the existence of an organizing theme or set of principles.

SAR 73: RECORDING AGREEMENTS AMONG MANY PARTIES

In engaging multiple parties in design, methodologies for recording ideas, proposals and decisions are vital. *SAR 65* developed methods by which Supports and Detachable Units (infill) could be produced independently. Subsequently, *SAR 73* (1974) extended such principles – together with the conviction that inhabitants should have a clearly defined role in complex planning processes – to the urban level. Based on observed self-organizing principles of historical urban areas, *SAR 73* offered a series of tools for both new areas and redevelopments (*12 Living Tissues*, 1975).

SAR 73 provides a set of graphic conventions for documenting agreements regarding morphology and function. In these conventions, morphology is further classified according to thematic and non-thematic buildings and spaces. It is by thematic forms and spaces that the main characteristics of an area are recognized. Non-thematic elements, while unusual, nevertheless appear in urban tissue in some regular way. The sum of all such documents jointly constitutes a tissue model: a way of conveying agreements concerning the placement and size of built form, space and functions in a neighborhood.

Horizontal and vertical positioning and dimensioning of buildings and open spaces are documented in a 'zoning' diagram that always includes the morphogy of thematic built and unbuilt areas. Functions or activities can be shown in the framework of the morphology or physical/spatial theme. Decisions concerning neighborhoods clearly involve many non-material factors – social issues, economics, individual preferences, etc. Ultimately, such vital considerations must also be reflected in tissue agreements.

The Tissue Method shifts the work of recording legal agreements concerning built environment to more explicit and information-intensive pictorial depiction (Tufte, 1990). In realized projects such as Beverwaard (1977), pictorial agreements constituted legal documents. Several decades later, many New Urbanist projects utilize a similarly graphic approach to recording agreements.

The Tissue Method defines building blocks of urban design in terms of elements (thematic and non-thematic space, form and activities); tissue models (graphically-conveyed sets of documents concerning form, space and activity); and plans (sited tissue models transformed to fit the characteristics of place).

Tissue models first position and dimension basic elements – house types, spatial configuration types (linear, courtyard, centralized, etc.) and functions (dwelling, shopping, meeting, etc.). Models are then adjusted to infill the actual site. Rational suboptimization of decision-making, a technique independently advocated by Christopher Alexander, follows: At each stage from the general to the particular, alternatives are discussed and firm agreements are established and recorded, prior to discussing next steps.

	M Morphology	**F** Function
Thematic Built Form	1	a 5 b
Thematic Open Space	2	6
Non-Thematic Built Form	3	7
Non-Thematic Open Space	4	8

a: concerns a document containing information on the position of functions

b: concerns a document containing information on the dimension of functions

Fig. B.1 Matrix for coding documents that describe an urban tissue. Drawing by Stephen Kendall, after *SAR 73*.

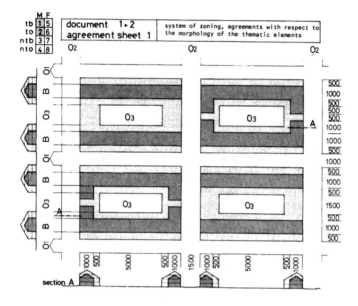

Fig. B.2 Tissue Documents 1 and 2 combined to form a tissue model. From *SAR 73*. Reprinted with permission.

ADDITIONAL READINGS ON THE SAR TISSUE METHOD

Habraken, N.J. (1964) The Tissue of the Town: Some Suggestions for further Scrutiny. *Forum.* **XVII** no. 1. pp. 22–37.

Habraken, N.J. *et al.* (1981) *The Grunsfeld Variations: A Report on the Thematic Development of an Urban Tissue.* MIT Department of Architecture, Cambridge, Mass.

Habraken, N.J. (1994) Cultivating the Field: About an Attitude when Making Architecture. *Places.* **9** no. 1. pp. 8–21.

Kendall, S. (1984) Teaching with Tissues, *Open House International.* **9** no. 4. pp. 15–22.

Reijenga, H. (1981) Town Planning Without Frills. *Open House.* **6** no. 4. pp. 10–20.

Reijenga, H. (1977) Beverwaard. *Open House.* **2** no. 4. pp. 2–9.

Stichting Architecten Research. (1975) *Living Tissues: An Investigation into the Tissue Characteristics of Twelve Residential Areas with the Aid of SAR 73.* SAR, Netherlands. Reprinted in *Open House International.*

Stichting Architecten Research. (1977) *Deciding on Density: An Investigation into High Density Allotment with a View to the Waldeck Area, The Hague.* Eindhoven, Netherlands.

Stichting Architecten Research. (n.d.) *Modellen en Plannen: de weefselmethode SAR 73 als hulpmiddel bij het stedebouwkundig ontwerpen.* Eindhoven, Netherlands.

Stichting Architecten Research. (1980) Neighborhood Improvement: A Methodological Approach. *Open House.* **5** no. 2. pp. 2–17.

Technische Hogeschool Delft. (1979) *Integratie van Deelplannen van Het Global Bestemmingsplan: Syllabus, van de M.M.V. de Stichting Architecten Research tot stand gekomen leergang.* September, Delft, Netherlands.

Appendix C

International Council for Research and Innovation in Building and Construction (CIB)

CIB is the international association providing a global network for international exchange and cooperation in research and innovation in building and construction. CIB supports improvements in building processes and in the performance of the built environment. The CIB program covers technical, economic, environmental, organizational and other aspects of the built environment during all stages of its life cycle. CIB addresses all steps in the process of basic and applied research, documentation and transfer of research results, and the implementation and actual application of them in practice.

Task Group 26 Open Building Implementation was formed November 1996. Our members are building owners, architects, interior designers, engineers, contractors, manufacturers, building economists and researchers in public and private organizations around the world. TG 26 studies and advocates developments toward an adaptable architecture for the 21st century. The mission of TG 26 is to document, stimulate and support implementation of Open Building in practice, and to disseminate the results of research aimed at improving Open Building. To realize this mission full participation is needed by professionals in many fields related to Open Building. The Task Group enthusiastically invites those interested to contact our co-ordinators and to search our web site: www.decco.nl/obi.

TASK GROUP (TG) 26
OPEN BUILDING
IMPLEMENTATION

Coordinators

Dekker, Karel
TNO Building and Construction
Research
Delft, Netherlands

Kendall, Stephen
Silver Spring, Maryland, USA

Members

Bao Jia-sheng
Southeast Univesity
Center for Open Building Research
and Development
Nanjing, China

Birtles, A.B.
The Steel Construction Institute
Ascot, UK

Boekholt, Jan Thijs
Eindhoven University of Technology
Eindhoven, Netherlands

Cuperus, Ype
Delft University of Technology
OBOM Research Group
Delft, Netherlands

Damen, A.A.J.
QD International BV
Rotterdam, Netherlands

Fukao, Seiichi
Tokyo Metropolitan University
Building Center of Japan
Tokyo, Japan

Geraedts, Rob
Delft University of Technology
Delft, Netherlands

Hankonen, Johanna
ARA – Housing Fund of Finland
Helsinki, Finland

Hermans, Marleen
KPMG Consulting
De Meern, Netherlands

Iwashita, Shigeaki
Atias Corporation
Tokyo, Japan

Kahri, Esko
RTS - Building Information Institute
Helsinki, Finland

Kamata, Kazuo
HUDc Housing Research Institute
Hachioji, Japan

Karni, Eyal
Technion – Israel Institute of
Technology
Haifa, Israel

Kiiras, J.
Helsinki University of Technology
Espoo, Finland

Kobata, Seiji
HUDc Design Division
Tokyo, Japan

Kobayashi, Hideki
Ministry of Construction Building
Research Institute
Tsukuba, Japan

Lahdenperä, Pertti
VTT Technology Research Center
of Finland
Tampere, Finland

Langelaan, J.W.R.
Langelaan Architects
Mississauga, Canada

Larsson, Nils
CANMET Natural Resources
Canada
Ottawa, Canada

Lee, T.K.
Architecture and Building Research
Institute
Ministry of Interior
Taipei, Taiwan

Lin, Li chu
National Kaohsiung University of
Science and Technology
Kaohsiung, Taiwan

Moseley, Richard
OBuild Consulting
London, UK

Murakami, Shin
Sugiyama Jogakuen University
Nagoya, Japan

Norton, Brian
University of Ulster at Jordanstown
Northern Ireland, UK

Okamoto, Shin
Building Center of Japan (BCJ)
CRICT-JARC
Tokyo, Japan

Olsen, Ib Steen
Ministry of Housing and Building
Copenhagen, Denmark

Pekkanen, Jukka
TEKES – Technology Development
Center
Helsinki, Finland

Salagnac, Jean-Luc
Centre Scientique et Technique du
Bâtiment (CSTB)
Paris, France

Sawada, Seiji
Tokyo, Japan

Scheublin, Frits
HBG - Hollandsche Beton Group BV
Rijswijk, Netherlands

Slaughter, Sarah
Massachusetts Institute of Technology
Cambridge, Massachusetts, US

Tanaka, Ryoju
Japan Association of General
Contractors
Tokyo, Japan

Teicher, Jonathan
American Institute of Architects
Washington, D.C., US

Tiuri, Ulpu
Helsinki University of Technology
Helsinki, Finland

Wang, Ming-Hung
National Cheng-Kung University
Tainan, Taiwan

Westra, Jan
Eindhoven University of Technology
Eindhoven, Netherlands

Yashiro, Tomonari
The University of Tokyo
Tokyo Japan

Guest members

Habraken, John
Emeritus Professor, MIT
Apeldoorn, Netherlands

Tatsumi, Kazuo
Emeritus Professor, Kyoto University
Kyoto, Japan

Utida, Yositika
Emeritus Professor, The University
of Tokyo
Tokyo, Japan

van Randen, Age
Emeritus Professor, Technical University
of Delft
Rotterdam, Netherlands

Glossary

BCJ Abbreviation of Building Center of Japan.

BL Abbreviation of Center for Better Living (Japan).

BRI Abbreviation of Building Research Institute (Japanese Ministry of Construction).

base building refers to the part of a multi-tenant building that directly serves and affects all tenants. In conventional North American practice, base buildings are constructed by speculative office building developers, leaving choice and responsibility for the remainder of the building to tenants during the fit-out phase. The base building normally includes the building's primary structure; the building envelope (roof and facade) in whole or part; public circulation and fire egress (lobbies, corridors, elevators and public stairs); and primary mechanical and supply systems (electricity, heating and air conditioning, telephone, water supply, drainage, gas, etc.) up to the point of contact with individual occupant spaces. Base buildings provide serviced space for occupancy; Supports are residential base buildings.

building knot is a term coined by OBOM to refer to the physical, decision-making and procedural entanglement inherent in conventional building processes.

Buyrent is the proprietary name for a new financial product in the Netherlands that provides legal, financial and management instruments for infill ownership. In purchasing the infill of their dwellings, tenants enjoy the same privileges and tax advantages as homeowners.

CHS. See **Century Housing System**.

CIB is the International Council for Research and Innovation in Building and Construction. Headquartered in Rotterdam, CIB is an international association providing a global network for international exchange and cooperation, supporting improvements in building processes and in the performance of the built environment.

capacity, in the context of Support/Infill building, refers to a range of variations in floor plan and use within the constraints of a given base building. More generally, capacity concerns the degree of Open

Building freedom offered by a higher level to a lower level.

Century Housing System refers to an Open Building approach to building design and construction developed in Japan by Professor Utida. CHS classifies and organizes the placement of building component systems based on modular coordination and the concept of durable years as it relates to each component group. Components with few durable years are installed after components with longer durable years.

comprehensive infill system. See **infill system**.

DIY is an abbreviation for Do-It-Yourself.

decision bundle refers to the totality of decisions under the control of a single party involved in the design, construction or mangement of buildings.

disentangling is a process of organizing technical systems and parties who control them such that a change of one system does not disturb (or only minimally disturbs) others.

durable years is a concept associated with a life-cycle accounting approach to building design, construction and management. Each subsystem is assigned an optimum expected length of useful life and is installed accordingly; parts having a relatively long life expectancy are installed first, followed by parts expected to have a shorter durable life.

decision cluster refers to a set of design, development, construction or other determinations or responsibilities appropriate to a single environmental level or entity. A Support is a decision cluster, as is infill.

detachable unit was the term first used to describe infill, that part of a residential multi-family building determined and controlled for the individual dwelling unit and preferably by the individual occupant. Among other places, the term occurs in *SAR 65*, a seminal report on the design of Supports.

environmental levels. See **levels**.

fit-out (tenant work) refers to the process or action of installing building infill, or to the physical products used in making habitable space in a base building. It may also modify or describe such processes or products. See **infill**.

fixed plan (fixed layout) refers to a dwelling plan arrangement that makes no specific provision for enabling subsequent transformation in response to user preferences.

HUDc. Abbreviation of Housing and Urban Development corporation (Japan).

infill (fit-out, tenant work, detachable unit) is the total configuration of physical parts determined for each individual occupancy – e.g., dwelling unit, office space or other tenancy – in the context of the higher level configuration or Support.

infill system refers to a specific selection of physical parts, having standardized interfaces and organized logistics, that can consequently be organized to suit a wide range of interior conditions and requirements. An infill system is ideally capable of being installed in any Support. Comprehensive infill systems may incorporate all components, subsystems and finishes, together with the design, cost estimating and logistics control software required to complete the work of fitting out a Support.

intervention describes the work of architects and other design and construction professionals. By implication, these professionals are viewed not as creators of environment, but rather as expert enablers or facilitators who help to realize the requirements and preferences of the many parties involved in environmental processes.

levels describe the interrelated configurations of physical elements and decision clusters that occur within a larger dependency hierarchy. In Open Building terms, the Support constitutes a higher level, while infill is a lower, dependent level: should the Support change, the infill is inevitably affected, although the infill can change without forcing change at the higher Support level. Environmental levels include: the urban (tissue) level; Support (base building or building) level; infill (fit-out) level; and furniture (furnishings) level.

MITI Abbreviation of Ministry of International Trade and Industry (Japan).

MOC Abbreviation of Ministry of Construction (Japan).

margin refers to an area of overlap belonging to two spatial planning zones, as when bay windows, porches and entryways extend the building's volume or facade into open space. A margin's size and features can be determined by a higher level, which may allow either little or great variation on the lower level.

OB Abbreviation of Open Building.

OBOM is the Open Building Simulation Model research and documentation group at the Technical University of Delft, in the Netherlands. The name was derived from the early use of simulation processes as a means of including industrial participants in the study of advanced technical solutions. The term 'Open Building' originated at OBOM.

open architecture broadly describes building design and construction practices which consciously create capacity for transformation.

open, openness describes the character of buildings – usually multi-tenant buildings – organized on levels to maximize capacity to transform while distributing choice, control and responsibility and reducing conflict during the process of change.

Open Building (OB) is the international movement based on organizing buildings and their technical and decision-making processes according to levels. In the West, Open Building was a partial successor to the Supports movement. Open Building is also a phrase used to describe projects, beliefs, methods or products which support such organizational principles.

ordering principles are rules of three-dimensional positioning. In Open Building, they serve to minimize interference among subsystems and define interfaces between them, thereby enabling separation of responsibilities and eliminating disruption.

PSSHAK is an acronym for Primary Support System and Housing Assembly Kit.

parcellation refers to the allotment or subdivision of available floor area within a Support.

plug-and-play is an electronics term that refers to products that can be safely installed without professionals and immediately used, like electronic consumer products. By implication, such products may subsequently be unplugged and removed or repositioned with equal ease. In the context of Open Building, plug-and-play refers to consumer-oriented infill products that attach to but are freely located within the Support.

plumbing tether refers to drainage performance requirements that effectively limit the positioning of plumbing fixtures away from vertical waste stack.

quality certified installer (QCI) An official designation enacted by an association comprising all utilities in the Netherlands for the purpose of enabling Open Building implementation. QCI certification permits an installer to submit a single certificate of completion for a multiple trade job. To be certified as a QCI, a company and its designated workers must demonstrate qualifications to install a specific infill 'product,' covering an approved scope of work.

residential Open Building is a multi-disciplinary approach to the design, financing, construction, fit-out and long-term management of buildings. It is based on the separation of base building and infill.

resource systems refers to supply systems, mechanical/electrical/plumbing systems or utilities.

reverberation is a rippling or echo effect in which construction or disruption in one system, part or level of a building disturbs others.

SAR (Stichting Architecten Research or **Foundation for Architects' Research).** The SAR was founded in the Netherlands in 1965 to 'stimulate industrialization in housing.' More generally, it sought to study issues surrounding the relationship between the architecture profession and the housing industry, and to chart new directions for architects in housing design.

S/I See **Support/Infill.**

shell is a term universally used to describe the exterior envelope of a building. In some contexts, it may also include the structural framework of a building.

social overhead capital is a term brought into Open Building usage in association with the Japanese Two Step Housing Supply System. It describes a Support characterized by high quality and long durability, specifically designed as public common property.

spaghetti effect is a term used by Van Randen to describe an entangled building condition in which unpredictable dependencies occur among parties involved, frequently leading to coordination breakdowns and lapses in quality control.

Support (Support structure) was a term first coined in John Habraken's book *Supports: An Alternative to Mass Housing.* It describes what might now alternatively be referred to as a residential base building, comprising the common part of a multi-tenant building.

Support/Infill refers to housing built according to Open Building separation of base building and fit-out.

Skeleton/Infill is a term used in Japan to describe the separation of building systems and decisions according to a subsystems approach distinguishing skeleton (including enclosure and most utility systems) from infill.

Supports broadly describes all or much of the constellation of ideas, principles, methods and technologies associated with the activities growing out of the early work of Habraken and the SAR, including the Support(s) Movement and Support(s) housing or principles.

supply systems refers to resource systems or utilities.

tartan band grid refers to the 10/20cm two-way band grid first developed by the SAR. It was subsequently adopted as a standard for modular coordination of building interiors throughout Europe. Variations on the band grid have been used in many other countries.

theme is a term shared by fields including music and design. It refers to recurring and easily recognized patterns of variable organization, as in 'theme and variation.'

thematic design is the design of variably recurring elements on any environmental level according to a set of organizing principles.

Tsukuba Method refers to a Japanese Open Building approach that employs the Two Step Housing Supply System and establishes a new system of property ownership and household control similar to a freehold arrangement.

Two Step Housing Supply System refers to a specific Japanese Open Building approach developed by Tatsumi at Kyoto University, with continuing development by Takada. It emphasizes the importance of a balance between public and private initiative in housing processes and advocates methods of housing design, construction and long-term management that clearly delineate community and individual household responsibilities.

unbundling refers to the sorting, separation and distribution to appropriate levels and parties of decisions concerning the use and placement of physical systems.

urban tissue refers to the environmental level normally associated with urban design. Tissue comprises coherent neighborhood morphology (open spaces, buildings) and functions (human activity). Neighborhoods exhibit recognizable patterns in the ordering of buildings, spaces and functions (themes), within which variation reinforces an organizing set of principles.

utilities refers to resource systems, mechanical/electrical/plumbing or supply systems.

variants are specific thematic variations of a typology or theme. The term also specifically refers to alternate unit plans for a given dwelling space in a Support.

vertical real estate refers to the valuation of a 'site' or allotment available to be fitted-out within a Support.

zero-slope drain line refers to greywater drain piping that is installed on a level surface. Such lines require no slope to drain, but carefully calculated relations between length from fixture to drainage manifold, pipe diameter and number of elbow fittings. Based on testing and demonstrated performance, zero-slope drains have been certified for installation as part of specific infill products in some jurisdictions.

Index

9 780367 398989